THE · OFFICIAL

ITV

Careers

H·A·N·D·B·O·O·K

Sue Davis – The ITV Association
in conjunction with
Monika Barnes – Thames Television
Don Cameron – ITN
Ray Hockey – Television South
Judith James – HTV
Mandy Patchett – Granada Television
John Paton – Scottish Television
Alison Todd – Granada Television

Headway · Hodder & Stoughton

British Library Cataloguing in Publication Data
Independent Television Association
  The ITV careers handbook.
  1. Great Britain. Television services. Career guides
  I. Title
  384'55'4'02341

  ISBN 0 340 50399 8

First published 1989
Second impression 1989

Typeset by Wearside Tradespools, Fulwell, Sunderland
Printed in Great Britain for Hodder and Stoughton Educational, a
division of Hodder and Stoughton Ltd, Mill Road, Dunton Green,
Sevenoaks, Kent by St Edmundsbury Press Ltd, Bury St Edmunds, Suffolk

CONTENTS

A MESSAGE FROM SIR HARRY SECOMBE

I've been involved in TV for more years than I'd care to remember, but they've also been my happiest years.

So here's some advice I'd like to offer:

If you want a great career then
Try TV
A career that gives great cheer
Is TV
It's a job you'll find exciting
Plus the prospects are inviting
Yes a job that you'll delight in
Is TV.

It's a job where you can proudly tell your wife
'I've got a job I know will be a job for life'
So here is what to do
To make sure you head the queue
Decide right now
Your future's in TV
Don't wait and see
Decide right now your future's in TV.

Harry Secombe

FOREWORD

Television is an expanding industry. The nature of Independent Television is changing and so are the opportunities for working within it. Clearly, people thinking about a career in ITV need to know what jobs there are and what qualifications and training are required for them. This book aims to provide such information in a way that will enable potential entrants to make decisions about the area of television which interests them most and to judge whether they are suited to it.

It is vital that ITV continues to attract the best talent from schools and colleges. I hope this book proves a first step in ensuring that it does.

Richard Dunn
Managing Director, Thames Television
Chairman, ITV Association Council

Part One

GENERAL INFORMATION

INTRODUCTION

This book has been designed to answer most of the questions that
people considering a career in ITV might wish to ask.

There is a description of each of the main jobs in television. Each
description gives an idea of the type of work that is done, the
conditions under which it is carried out, and the type of person who
is best suited to the work. There is also a 'Typical Recruitment
Profile' wherever possible. This describes the qualifications, experience and personality, etc., that are required for the job. In some cases,
however, the backgrounds of people in the job are so varied that it is
not possible to describe a typical recruit.

The book also contains a number of sections of a more general
nature which should provide useful additional information for people
who are planning to apply for jobs.

The book is updated approximately every three years to ensure that
it keeps pace with changes in the industry. This is the third edition.

Points to Note

1 All jobs are open to both men and women, regardless of marital
   status or ethnic origin. A disability or health problem does not
   preclude full consideration for a job, and applications from
   disabled people are welcome.
2 There are slight variations in the qualifications required by
   different companies to do the same job. However, anyone
   matching the 'Typical Recruitment Profile' should be in a position
   to apply for a job in any of the ITV companies.

   The recruitment profiles list the academic qualifications that are
   needed for various jobs. Equivalent qualifications are of course
   acceptable.
3 There are also variations in the content of jobs between the ITV
   companies, and the descriptions can therefore only give a broad
   picture. Many of the variations are described in the text. Most
   arise from differences in the balance of programmes made by each
   company, while others may be organisational differences. The
   larger ITV companies spend a relatively high proportion of their
   time making dramas and light entertainment programmes, etc.,
   which will call for regular use of a wide range of skills in areas

1

such as make-up, wardrobe and special effects.

The smaller companies spend a smaller proportion of their time on this type of programme. Staff are, however, increasingly called upon to use specialised skills. Staff in smaller companies may also be responsible for a wider range of duties than in larger companies, and may sometimes find that they take responsibility at an earlier stage.

4 The changing nature of the television industry means that the dividing line between many jobs is becoming increasingly blurred. Employees are being called upon to learn a wider range of skills. For example, camera operations is in some instances combined with sound, editing and lighting. Most employees of the future will be 'multi-skilled'. They will perhaps have in-depth knowledge and experience in one or two functions, and basic training in other areas to enable them to assist where necessary. The job definitions in this book should therefore be taken as a guide only. The ways in which various jobs are combining varies from company to company.

5 No mention is made of salaries as they change regularly, and vary considerably with overtime, shift allowances, etc. Salaries are, however, highly competitive for all grades of staff.

6 Jobs in television fall broadly into the following categories: *Technical* such as Engineering; *Creative* such as Director, Set Designer, Make-up Artist; a *Mixture* of both such as Camera Operator, Editor, Sound Technician or Vision Mixer; *Journalistic* such as Researcher or Journalist; or *Administrative*. There is obviously overlap between these categories, for example a Director must do some administration, and a Make-up Artist must be aware of how the properties of the camera will affect the final appearance of a made-up Performer. In general, however, would-be employees should decide in which area their abilities lie before applying for a post in television.

7 With very few exceptions, there is no set career path to any job in television and it is difficult to give step-by-step guidance on how to get a job. Academic qualifications are important of course, and since competition for posts is tough it is often the applicant who has more qualifications to offer who gets the job. The exact subject that has been studied may not, however, be relevant. The main qualification that is needed is a flair for the job which can be demonstrated either through previous professional experience or through hobbies and pastimes.

8 There are a number of jobs which are not usually available to outside applicants unless they have previous experience. These include Producer, Director, Lighting Director, and Floor Manager. When trainee vacancies occur in these jobs, they are advertised internally and may be filled by staff from any department.

Trainee vacancies for jobs in other areas are also advertised internally as well as externally.

9 The number of vacancies which are filled by external applicants is comparatively small and competition is tough. In 1988, a total of 1103 people were recruited by the ITV companies; 6 of these came directly from school, 55 from college or university, and 588 joined from other industries. The remainder were moving from one company to another from within the television, film, theatre and radio industries, or from freelance to full-time employment. Of these external applicants, 381 were appointed to posts in administrative, clerical and secretarial functions, 230 to sales, and 102 to engineering. Smaller numbers were appointed to all other posts including those in programme-making. The figures for 1988 are fairly typical of recent years. The future is less certain. Recruitment is unlikely to rise significantly in the short term, and it may indeed fall. Opportunities do occur, however, and it is the applicants with determination as well as talent and qualifications who are usually successful.

10 Although there are no fixed age limits for applicants, most successful candidates for trainee posts in all areas tend to be in their twenties. Younger applicants may apply for craft and electrical apprenticeships or junior clerical posts while older applicants are frequently successful in applying for administrative, clerical and managerial posts. They are also welcome in engineering posts. The over-riding factor is, however, a flair for the job and a highly talented applicant of any age can be accepted for any job.

11 Staff who are employed in any aspect of programme-making, and in many administrative and managerial jobs, must be prepared to work at any time of the day or week, and in any place. Television is not a nine-to-five job, so it can be disrupting to social and personal life. Potential employees should be aware of this before applying for posts.

12 ITV recognises four trade unions which represent staff in different functions. It is not necessary to be a member of a trade union before applying for a post on the staff of an ITV company. You

will have a free choice of whether or not to join the appropriate union after appointment.

HOW TO APPLY FOR JOBS IN INDEPENDENT TELEVISION

The addresses of the ITV companies are given below. Addresses marked with an asterisk (*) indicate the location of the main television studios. Other locations usually employ small numbers of sales (but not marketing) and administrative staff. Enquiries about vacancies in these subsidiary offices should be sent to the company's studio address.

You should write to the Personnel Department at the company of your choice only if you have the qualifications necessary for the job you wish to do (or in the case of academic qualifications, if you expect to gain the qualification in the near future). The company will tell you if there are any current vacancies in the job you wish to apply for. General lists of vacancies are not available. Suitable applications may be held on file for a short period, but not usually for more than three months.

The ITV companies receive a very large number of enquiries, and in order to ease the pressure on Personnel Departments, we would prefer you not to telephone if possible.

Personnel Departments are happy to provide careers advice.

Anglia Television

Personnel Officer
Anglia Television Ltd*
Anglia House
NORWICH NR1 3JG
Tel: 0603 615151
Other offices in London,
Manchester,
Chelmsford, Luton, Ipswich,
Northampton and
Peterborough.

Border Television

Personnel Manager
Border Television plc*
Television Centre
CARLISLE CA1 3NT
Tel: 0228 25101
Other offices in London.

Central Independent Television

Recruitment & Training Officer
Central Independent Television
plc*
Central House
Broad Street
BIRMINGHAM B1 2JP
Tel: 021 643 9898

Recruitment & Training Officer
Central Independent Television
plc*
East Midlands Television
Centre
Lenton Lane
NOTTINGHAM NG7 2NA
Tel: 0602 863322
Other offices in London and
Oxford.

Channel 4

The programmes for Channel 4
are made by the ITV companies
and by independent producers.
Channel 4 does not therefore
recruit programme-making
staff. Sales of advertising time
are also handled by the ITV
companies at the time of going
to press. Correspondence
should be addressed to:
Channel 4 Television Company
Ltd
60 Charlotte Street
LONDON W1P 2AX
Tel: 01 631 4444

Channel Television

Secretary to the Managing
Director
Channel Television*
The Television Centre
St Helier
Jersey
CHANNEL ISLANDS
Tel: Jersey 0534 73999
Other offices in Guernsey.
Note: Channel Television is
unable to offer employment to
persons who are not qualified as
Channel Islands residents under
the Jersey and Guernsey
housing regulations.

Grampian Television

Personnel Officer
Grampian Television plc*
Queen's Cross
ABERDEEN AB9 2XJ
Tel: 0224 646464
Other offices in London,
Dundee, Edinburgh and
Inverness.

Granada Television

Personnel Manager
Granada Television Ltd*
Quay Street
MANCHESTER M60 9EA
Tel: 061 832 7211

Personnel Officer
Granada Group Ltd
36 Golden Square
LONDON W1R 4AH
Tel: 01 734 8080
Other offices in Liverpool.

▨ HTV

Personnel Manager
HTV Wales*
The Television Centre
Culverhouse Cross
CARDIFF CF5 6XJ
Tel: 0222 590590

HTV West*
Television Centre
Bath Road
BRISTOL BS4 3HG
Tel: 0272 778366

HTV Wales*
Television Centre
MOLD
Clwyd CH7 1YA
Tel: 0352 55331
Other offices in London.
Note: All applications for jobs
should be sent to Cardiff.

▨ London Weekend Television

Personnel Manager
London Weekend Television
Ltd*
South Bank Television Centre
Kent House
Upper Ground
LONDON SE1 9LT
Tel: 01 261 3434
Other offices in Manchester.

▨ S4C (Welsh Fourth Channel)

Personnel Officer
S4C
Sophia Close
CARDIFF CF1 9XY
Tel: 0222 343421
Note: S4C does not employ
programme-making staff.

▨ Independent Television News

Manager, Personnel and
Industrial Relations
Independent Television News
Ltd*
ITN House
48 Wells Street
LONDON W1P 4DE
Tel: 01 637 2424

▨ Scottish Television

Recruitment and Training
Manager
Scottish Television plc*
Cowcaddens
GLASGOW G2 3PR
Tel: 041 332 9999
Other offices in London,
Edinburgh and Manchester.

TSW (Television South West)

Personnel Manager
TSW (Television South West Ltd)*
Derry's Cross
PLYMOUTH PL1 2SP
Tel: 0752 663322
Other offices in London, Bristol and Yeovil.

Thames Television

Personnel Manager
Thames Television Ltd*
Teddington Studios
Teddington Lock
TEDDINGTON
Middlesex TW11 9NT
Tel: 01 977 3252

Personnel Manager
Thames Television Ltd*
Thames Television House
306–316 Euston Road
LONDON NW1 3BB
Tel: 01 387 9494
Other offices in Birmingham.

TV-am

Personnel Administrator
TV-am*
Breakfast Television Centre
Hawley Crescent
LONDON NW1 8EF
Tel: 01 267 4300

TVS (Television South)

Personnel and Recruitment Manager
TVS*
Television Centre
SOUTHAMPTON SO9 5HZ
Tel: 0703 634211

Personnel Manager
TVS*
Television Centre
Vinters Park
MAIDSTONE ME14 5NZ
Tel: 0622 691111
Other offices in London, Reading, Dorchester, Poole, Brighton, Manchester.

Tyne Tees Television

Personnel Manager
Tyne Tees Television Ltd*
The Television Centre
City Road
NEWCASTLE UPON TYNE
NE1 2AL
Tel: 091 261 0181
Other offices in Middlesbrough and Manchester.

Ulster Television

Personnel Manager
Ulster Television Ltd*
Havelock House
Ormeau Road
BELFAST BT7 1EB
Tel: 0232 328122
Other offices in London.

Yorkshire Television

Personnel Executive
Yorkshire Television Ltd*
The Television Centre
LEEDS LS3 1JS
Tel: 0532 438283
Other offices in London,
Sheffield, Hull, Ripon,
Lincoln and Grimsby.

Independent Broadcasting Authority

70 Brompton Road
LONDON SW3 1EY
Tel: 01 584 7011

Note: The IBA (Independent
Broadcasting Authority)
employs mostly administrative
and engineering staff. It does
not employ programme-making
staff.

WHERE TO LOOK FOR JOB ADVERTISEMENTS

Some publications which carry job advertisements for ITV are: *The
Guardian* Media Page, *Broadcast*, *Campaign* (Sales and Marketing
posts), *Sight and Sound*, *Stage and Television Today*, *Audio Visual*
(technical posts), *UK Press Gazette*, and appropriate technical and
hobby publications.

Advertisements are also placed in the national and local press.

Jobs for engineers are advertised in a wide range of professional
and technical journals.

Many of the specialist publications are available in newsagents in
London's West End, but elsewhere they may only be available on
special order through newsagents, or at local libraries.

THE ORGANISATION OF ITV

The Independent Television (ITV) regional programme companies are 15 totally separate companies, each under different ownership, but working together to produce a coordinated output of programmes for ITV throughout the United Kingdom.

Each company is responsible for a particular geographical region, although there are some areas in which the output of more than one company can be received. The regions are:

- *Anglia Television* – East of England
- *Border Television* – The border of England and Scotland, and the Isle of Man
- *Central Independent Television* – East and West Midlands
- *Channel Television* – Channel Islands
- *Grampian Television* – Northern Scotland
- *Granada Television* – North-West England
- *HTV* – Wales and the West of England
- *London Weekend Television* – The London area
- *Scottish Television* – Central Scotland
- *Television South West* – South-West England
- *Thames Television* – The London area
- *TVS* – South and South-East England
- *Tyne Tees Television* – North-East England and North Yorkshire
- *Ulster Television* – Northern Ireland
- *Yorkshire Television* – Yorkshire and Humberside

The London area is served by two companies, Thames Television which is responsible for weekday transmissions, and London Weekend Television which takes over from Friday night to Sunday night. Since London is the largest centre of population in the UK, there is currently a sufficiently large market for advertising time to sustain two companies.

The ITV companies vary considerably in both size and the advertising revenue they receive. Thames Television, for example, currently employs approximately 2000 staff while Channel Television employs about 100. About 15 000 staff are employed in total, but the number is falling.

The larger companies tend to produce a higher proportion of the more expensive programmes such as dramas and light entertainment. There is therefore greater scope for the use of specialist skills such as Make-up, Costume and Set Design in these companies.

All companies lay great stress on programmes produced for their own regions which are predominantly news and current affairs, and documentaries. The regional nature of ITV is one of its great strengths and the 1988 White Paper on Broadcasting emphasised that it should continue to be so.

The main source of income to the ITV companies is from the sale of advertising time, although some income is derived from the sale of programmes to other television stations around the world, and to the video market. ITV differs from most other businesses in several ways. Although the quality of programmes is a strong determinant of the sale of advertising time, the sale of the main product (i.e. programmes) is not the main source of income.

Another respect in which ITV differs from most other businesses is that the programme companies must compete for the renewal of their franchises to operate every few years.

How the ITV Network Operates

The network (i.e. national) schedule on ITV is put together by a committee consisting of the Programme Controllers of seven of the ITV companies. They are assisted in this by the Programme Planning Secretariat at the ITV Association in London. The proposed schedule is placed for approval before the Network Programme Committee which consists of the Managing Directors and Programme Controllers of some of the companies.

There are certain fixed points in the schedule (e.g. 'News at Ten', and 'Coronation Street' at 7.30 p.m. on Mondays and Wednesdays). These could, however, be changed if the Network Programme Committee thought a better schedule would result.

During peak viewing hours the programmes are the same throughout the country, but early and late in the evening, each region provides locally transmitted programmes. Each company also decides its own night-time schedule. The operation of the network is a complex business calling for coordination, cooperation, and accurate timing. Each programme must last for exactly the right duration for the slot in which it is to be transmitted. In simple terms the majority of networked programmes are transmitted from the company which owns them. 'News at Ten' for example is transmitted from the ITN studios to the regional ITV companies and then via the local transmitters to homes around the country. Advertisements are transmitted regionally, so the length and timing of commercial breaks is crucial if all companies are to leave and rejoin the network at the right time.

Some of the programmes which are shown on ITV are made by the ITV companies themselves. Some, such as feature films and American series are purchased from outside, and some are commissioned by the ITV companies from independent producers. If an ITV company either makes or commissions a programme which is shown on the network, all participating companies will contribute to the cost in accordance with a formula based on their proportionate advertising revenue.

Independent Television News (ITN)

ITN is currently owned by the ITV companies. It has its own studios and employs a wide range of staff including many journalists. It provides national and international news programmes and bulletins around the clock for ITV, such as 'News at One', 'News at 5.45' and 'News at Ten'. It also provides Channel 4 News, and other news-based programmes for ITV, Channel 4 and other customers. With its excellent coverage of events throughout the world, ITN has an international reputation for reporting news with accuracy and authority.

Channel 4 Television

Channel 4 began transmitting in 1982, covering all regions of Britain except Wales. Its brief is to provide distinctive and innovative programming, particularly in areas of programming which are not catered for by ITV. It also plans its schedule to complement that of ITV where possible.

Channel 4 commissions its programmes from the ITV companies and independent producers. The channel is currently financed by subscription from the ITV companies who in return have the right to sell Channel 4 advertising time in their regions. This may change as a result of the 1988 White Paper on Broadcasting and Channel 4 may sever its financial connections with the ITV companies.

Since Channel 4 does not make its own programmes, the range of jobs is somewhat limited. There are no opportunities for Directors,

Production Assistants, Camera Operators and others involved in the making of programmes. There are ten main departments in the company which are – Engineering, Presentation, Information Systems, Commissioning, Marketing and Press, Programme Sales, Programme Planning, Finance, the Legal Department, and Personnel and Administration.

S4C

The Welsh Channel 4 Authority was established as an independent authority under the Broadcasting Act of 1981 with responsibility for a service of Welsh and English programmes on the fourth channel in Wales. The service, known as S4C, consists of about 28 hours a week in peak time of Welsh language programmes and more than 60 hours English language output from Channel 4. The Welsh programmes are provided by HTV, the BBC, and independent producers. The service is broadcast on the IBA's fourth channel transmitter network in Wales and is funded by the IBA from the ITV companies' fourth channel subscriptions.

TV-am

TV-am holds the Independent Television franchise for breakfast-time on ITV. Like Channel 4, TV-am covers the entire country, but unlike Channel 4 it makes its own programmes and sells its own advertising time. It is an independently owned company.

The Independent Television Association

The ITV Association is the central secretariat of the companies. It employs about 120 staff who service the committees which formulate industry policy on finance, marketing, industrial relations, training and engineering. It also coordinates the work of the Programme Controllers who compile the network schedules, and has a separate department which ensures that television advertisements meet the standards laid down by the Independent Broadcasting Authority.

The Independent Broadcasting Authority (IBA)

The IBA is currently the body set up by Parliament to provide public independent television and radio services via contracts with the programme companies. Under the Broadcasting Act 1981, the IBA has four main functions in respect of Independent Television.

1 To select ITV programme companies

The 15 regional companies hold fixed-term contracts or franchises with the IBA to provide the ITV programme service in their regions. TV-am holds a similar franchise for the nationwide breakfast-time service.

At the end of the contract period, the companies must bid for the renewal of their franchise in competition with other contenders. The IBA decides to whom the contracts should be awarded. The decision is based mainly on the ability of applicants to offer a regional television service of a high standard. The next franchise renewal will take place in 1992.

2 To supervise programme planning

The IBA is responsible for ensuring that ITV programmes provide a balance of information, education and entertainment at a high standard. It must also ensure that programmes are accurate, impartial and are not offensive to good taste and decency.

3 To control advertising

The IBA ensures that there is a clear distinction between the advertisements and the programmes, and that advertisements comply with the regulations to be, for example, legal, decent and honest. The frequency and amount of advertising is also strictly controlled by the IBA; for example no more than seven minutes of advertisements in non-peak time and seven and a half in peak time (18.00–23.00 hrs) are currently allowed in any clock hour.

4 To transmit programmes

The IBA currently owns and operates all ITV and Channel 4 transmitters and its engineering staff are responsible for ensuring that the pictures and sound are of a high technical quality when they reach the homes of the general public.

THE FUTURE OF ITV

In November 1988, the Government published a White Paper 'Broadcasting in the 90s: Competition, Choice and Quality'. This set out its plans for the future of television. It contained a number of proposals designed to increase diversity in broadcasting. These included:

- A fifth terrestrial (non-satellite) television channel funded by advertising, sponsorship and subscription.
- The abolition of the IBA and its replacement by a new Independent Television Commission which will reflect the changing nature of the industry. The new body will regulate cable, satellite and commercial terrestrial television with a 'lighter touch'.
- The awarding of ITV franchises by competitive tender. Applicants will first have to satisfy the Independent Television Commission that they can meet a 'quality threshold' in respect of their programmes, and then the franchises will be awarded to the highest bidders.
- The relaxation of rules requiring the ITV companies to provide a balanced output of programmes. They will be obliged to produce a 'diverse programme service', but the Government's view is that the proliferation of television channels makes it unnecessary for one channel to be obliged to meet all needs.
- A minimum of 25 per cent of original programming on ITV must be made by independent producers rather than the ITV companies themselves.
- The ITV companies may choose, under the new franchise arrangement, not to make any programmes themselves but to commission them all from independent producers or other organisations if they wish.

The proposals contained in the White Paper will, if accepted, take effect from 1992, but the ITV companies are already examining how they will respond to this new environment.

The White Paper was prompted in part by the need for the Government to re-examine its policies on broadcasting in the light of changes in technology which have made possible the proliferation of satellite and cable television channels. Until recently, most viewers in the UK were able to receive four television channels, but by 1990 many channels will be available following the launch of satellite services such as Astra, BSB and Eutelsat II. These channels will be funded by a mixture of advertising and subscriptions from the viewing public. For the first time, ITV will face stiff competition for television advertisers.

Recent changes in technology have also made possible changes in methods of programme production. The introduction of lightweight electronic cameras which combine sound and vision recording means that one person is able to carry out tasks which were previously done by two or more.

All of these factors will have an influence on the employment market in the future. There is likely to be a shift in balance away from full-time employment in the major broadcasting companies towards freelance or short-term contract employment, either with the broadcast companies, or with independent production companies and facility houses. The nature of jobs is also likely to change, with more people becoming multi-skilled.

There is currently much speculation about the future. What is certain is that the industry faces a time of change in the way that it will be run, in the technology it uses, and in its relationships with other broadcasters. It will be an exciting time where new ideas will flourish, and where new opportunities will abound.

OTHER EMPLOYMENT OPPORTUNITIES IN TELEVISION AND RELATED AREAS

The pattern of employment in television in the UK is changing rapidly and there are increasing numbers of opportunities outside the mainstream broadcasting companies. Those seeking a career in television should not therefore limit their ambitions to ITV and the BBC. In the past, the majority of people in the industry have been employed by ITV and the BBC; however, a number of factors have contributed to a dramatic change. First, the number of television stations available to the public has increased enormously and this has opened up new opportunities for programme makers. Secondly, the government now expects the BBC and ITV to achieve a target of commissioning 25 per cent of the programmes for their own output, not from their own staff, but from independent production companies. There is therefore a shift away from full-time employment in major broadcasting companies towards freelance employment and short-term contract employment in the industry as a whole.

If you are unsuccessful in your application for employment in ITV, you may wish to consider the following alternatives.

BBC Television

The BBC is, of course, a large employer of television personnel. Competition for posts, however, is as fierce as it is in ITV. The nature of jobs and the qualifications required may differ between the BBC and ITV and you should therefore make separate enquiries before applying for posts.

The address to which you should apply is:
BBC Corporate Recruitment Services
5 Portland Place
LONDON W1A 1AA.

Independent Production Companies

Independent production companies, or 'independents', should not be confused with Independent Television (ITV). Their role is to make a wide range of programmes and other visual products, but they do not transmit them to the public. Independents may be commissioned by ITV, Channel 4, the BBC, or a cable or satellite station to make a particular kind of programme that will fit into their schedules. They may also make television commercials, or sponsored videos on subjects such as health or road safety. In addition they may be commissioned by industry to make videos which promote a product. Some make videos for training purposes; indeed the range of activities is endless. Some specialise in a particular field, and some provide a range of services.

The size of such companies varies considerably. A few are fairly large but the majority are small. Many consist of just one creative person who hires in crew and facilities for making the production as needed. Very few independents employ people on a full-time basis. Many former employees of the BBC and ITV have established their own independent companies and the market is full of talented and experienced people who can devise a format for a programme and plan how it is to be made. That does not mean there are no opportunities for newcomers, but it is a business where many companies cease trading within a short time. It is undoubtedly an advantage to have some prior knowledge of the television industry, how programmes are made, and how the system operates before entering this field. Further details on independent companies can be obtained from the Association of Independent Producers, 17 Pulteney Street, London W1.

Facility Houses

Facility houses (or facility companies) generally provide a service to the same markets as independent production companies. Their role is primarily to provide technical facilities and staff to operate them. Some offer studio facilities including cameras, sound and vision mixing. Some offer facilities such as special effects or the duplication of tapes. They may be hired by independent producers who do not have their own facilities on 'dry hire' (facilities only) or 'wet hire' (facilities plus operational crew).

They may also be used by broadcast companies when they do not have the necessary facilities in-house. Most facility houses employ a small number of administrative staff as well as engineers and operational crews. They may also employ freelance crews and technicians from time to time.

There are generally no set career paths to follow into these branches of the industry, but some previous experience is an advantage. There are also opportunities for young people who can demonstrate genuine interest and talent, and perhaps have some appropriate qualifications but no professional experience.

Lists of facility houses and independent production companies can be found in *Kemps International Film and Television Year Book*, the

Broadcast Production Guide (a supplement produced by *Broadcast* magazine), the *Audio Visual Directory* and *Yellow Pages*.

With the exception of *Yellow Pages*, all of these publications are expensive to buy and you are therefore advised to enquire at your local library.

Job vacancies in production companies and facility houses are normally advertised in *Broadcast* or other trade papers, or the media page of *The Guardian* newspaper.

Educational Television

Television is an important medium for teaching at all levels in education. Many universities, colleges and even schools have their own television studios for making programmes. The standard of equipment used for educational television varies enormously from equipment suitable for domestic use right up to broadcast standard. There are usually only small numbers of staff employed in these units. Most are technical staff, but specialists in educational psychology and learning theory (often called media resources officers) are also employed to help user departments to gain maximum benefit from television.

Industrial Television

Many large industrial companies, and organisations such as the police and Ministry of Defence, have recognised the benefits of closed-circuit television (CCTV) as a means of communicating with their staff. Some have television facilities for staff-training, conferences, promotions, etc. Only a small number of staff are employed and most of them are technicians.

Equipment Manufacturers

The many companies which manufacture equipment for use in television studios offer opportunities for electronics engineers to work on the development and maintenance of cameras, video tape recorders, sound consoles, vision mixers, etc.

Independent Broadcasting Authority (IBA)

The main offices of the IBA are in London where mainly administrative and managerial staff are employed. The IBA also employs engineering staff, mainly at Winchester and at regional bases. Regional Officers liaise with the ITV companies to ensure that the quality of the sound and picture transmission is maintained. Other engineers are employed to maintain the many transmitters around the country.

The IBA has a training scheme for transmitter engineers and other agencies requiring technical training, based at its training college at Seaton in Devon.

Further information can be obtained from:
The Independent Broadcasting Authority
70 Brompton Road
LONDON SW3 1EY
Tel: 01 584 7011

There are, of course, opportunities in other areas of the media, such as the film industry, radio, and print journalism.

The Film Industry

In recent years the British film industry has enjoyed a worldwide reputation for excellence. In addition to home-grown productions, many foreign film makers consider that the technical and production facilities offered in the UK are the best available. The *Star Wars* and *Superman* films, for example, were made by Hollywood producers using British technicians at British studios.

The permanent staff at studios such as Pinewood and Shepperton are mainly administrative, or are employed in crafts such as carpentry and upholstery. Film producers generally hire the studio and bring with them a freelance team of film camera operators, sound recordists, make-up artists, etc.

Entry into film production is generally at a very junior level, perhaps as a 'runner', with opportunities for progression for those with talent.

The National Film and Television School runs courses at post-graduate level for people with a demonstrable talent for writing,

directing, producing, or photography. Details can be obtained from:
NFTS
Beaconsfield Studios
Station Road
BEACONSFIELD, BUCKS HP9 1LG

Jobfit

Jobfit is a two-year basic training scheme for technical and production grades in the film industry, and is jointly run by the employers
and the trade union (ACTT).

Up to fifty traineeships a year are offered, and trainees are attached
to a variety of film productions around the UK to gain experience in
pre-production, production and post-production. After one year
they specialise in one particular job such as cameras, sound or editing.
On completion of training, trainees join the freelance job market
in junior film technician grades. No particular qualifications are
specified and there is no rigid age limit for applicants.

For details, write to:
The Administrator
JOBFIT
Fourth Floor, 5 Dean Street
LONDON W1V 5RN
Tel: 01 734 5141

CYFLE

CYFLE was set up to meet the need for Welsh-speaking technicians
in the freelance film and television sector in Wales. Training is in the
form of an apprenticeship where trainees are attached to various
productions over a two-year period. The ability to speak Welsh is
essential, or if not proficient in the Welsh language applicants will
have shown the initiative to learn. This scheme is sponsored by S4C,
TAC (Welsh Independent Producers), and the trade unions ACTT
and BETA.

Radio

There are Independent Local Radio (ILR) stations in most parts of the country. They provide opportunities for a wide variety of people, including Engineers and Technicians, Producers, Directors, Editors, Journalists, Presenters and sales and administrative staff.

Details of jobs can be obtained from your local radio station or from the Association of Independent Radio Contractors at 46 Westbourne Grove, London W2 5SH.

The BBC also has a network of local radio stations in addition to its national radio services.

The Joint Advisory Council for the Training of Radio Journalists (JACTRJ) approves full-time courses in radio journalism at selected colleges (see 'Pre-Entry Education and Training Courses'). JACTRJ consists of representatives from Independent Local Radio and BBC Radio, the National Union of Journalists, and educational establishments which run radio journalism courses.

Print Journalism

There are opportunities for journalists not only in the Fleet Street newspapers but also on local newspapers and on all types of magazines. Some organisations offer their own training schemes and a number of further education colleges around the UK offer recognised courses in print journalism.

Details of careers in regional and local newspapers (but not nationals) can be obtained from:

The Training Department
The Newspaper Society
Bloomsbury House, Bloomsbury Square
74–77 Great Russell Street
LONDON WC1B 2DA

Details of job vacancies can be obtained from your local or national newspaper or magazine.

PRE-ENTRY EDUCATION AND TRAINING COURSES

There are many courses throughout the country which offer education and training in television. Some are run by the public sector (in colleges, polytechnics and universities), and some are run by private organisations. Some offer practical vocational training, while others offer academic study of the media. The choice is bewildering and it can be very difficult for someone wishing to find employment in television to know which one to choose.

ITV does not officially recognise any courses, mainly because the quality of a course can change overnight if a key tutor leaves, or certain equipment is no longer available. There are, however, a number of courses which are known to the ITV companies and are thought to be of a good standard at the time of going to print. These are listed below. Employment cannot be guaranteed on completion of any course, and inclusion in the list does not necessarily mean that the ITV companies recruit from the course.

Courses at GCE 'A' Level or Similar

Address	Course title	Length of course	Subject area
Bournemouth & Poole College of Art & Design Wallisdown POOLE BH12 5HH Tel: 0202 533011	BTEC National Diploma in Audio Visual Studies	2 years	Practical self-exploratory course covering audio visual design. Options in video, sound and slide tape programming.
Gwent College of Higher Education Clarence Place Newport GWENT NP9 0UW Tel: 0633 59984	BTEC HND in Film and TV Practice	2 years	Film and TV theory and practice, Animation, Computer Graphics, etc.

Address	Course title	Length of course	Subject area
Sandwell College of Further & Higher Education Wednesbury Campus Woden Road South Sandwell, WEST MIDLANDS WS10 0PE Tel: 021 556 6000	Certificate in TV and Audio Production (to include City and Guilds 110 Media Techniques, TV and Video Competences)	1 year	TV Production. Design for TV, Sound Production for Radio, TV and Audio Presentation, Technical Operations, Word Processing for Autocue, Media Studies, Computers for Graphics Generation.
West Kent College of Further Education Brook Street TONBRIDGE 2PW TN9 Tel: 0732 358101	BTEC National Diploma in Design (Communication: Media Arts Technology)	2 years	Broadcasting, Video, Sound, Photography, and Film Making, Graphic Design, Theatrical Design, Audio Visual Design, Information Technology Design, Desk-top Publishing, etc.

Courses at First Degree Level or Similar

Address	Course title	Length of course	Subject area
Bournemouth & Poole College of Art & Design Wallisdown POOLE BH12 5HH Tel: 0202 533011	(a) BTEC HND in Photography	2 years	Practical, self-exploratory course covering Photography in Advertising/Fashion; Editorial.
	(b) BTEC HND in Film and TV Studies	2 years	Practical, self-exploratory course covering Film and TV.
Christ Church College, CANTERBURY CT1 1QU Tel: 0227 762444	BA/BSc. Hons in: (a) Radio, Film and Television Studies	3 years	The Media – theory and practice; in combination with *one* of the following: Art, English, Geography, History, Information Technology, Mathematics, Movement Studies, Music, Religious Studies or Science.
	(b) Music with Film and TV Studies	3 years	Musical performance, Musicianship, Acoustics, Electronics, Music in the

Address	Course title	Length of course	Subject area
			Media, etc. Researching, writing and production techniques.
Harrow College of Higher Education Northwick Park HARROW Tel: 01 864 5422	BA Hons in Photography, Film and Video	3 years	Students study all aspects of film, video and photography in the first year and may or may not choose to specialise in any one area in years 2 and 3.
Kings Alfred's College Sparkford Road WINCHEST-ER Tel: 0962 841515	BA Hons in Drama, Theatre and Television Studies		
London College of Fashion 20 John Princes Street LONDON W1M 0BJ Tel: 01 629 9401	BTEC Higher National Diploma in Theatrical Studies Option A, Specialised Make-up	2 years	The specialist skills of the TV Make-up Artist included as part of a wider course.

Address	Course title	Length of course	Subject area
London College of Printing Elephant & Castle LONDON SE1 6SB Tel: 01 735 8484	(a) BA Hons Photography	3 years	Photography.
	(b) BA Hons Film and Video	3 years	Film and Video.
Manchester Polytechnic Capital Building School Lane Didsbury MANCHEST-ER M20 0HT Tel: 061 434 3331	BA Hons in Design for the Communica-tions Media	3 years	Film and TV Production and Design.
North Cheshire College Padgate Campus Fearnhed WARRING-TON Tel: 0925 814343	BA Hons in Media/ Business Management	3 years	Analysis of the media in society. Sound production, Photography, and Video.

Address	Course title	Length of course	Subject area
North East London Polytechnic Greengate House Greengate Street LONDON E13 0BG Tel: 01 590 7722	BA Hons in Fine Art	3 years	Photography, Film and Audio are included in a wide range of fine arts activities.
Plymouth College of Art & Design Tavistock Place PLYMOUTH PL4 8AT Tel: 0752 264774	BTEC, HND in Design (Photography, Film and TV)	2 years	Film and TV practice and theory. An emphasis on practical experience.
Plymouth Polytechnic Drake Circus PLYMOUTH PL4 8AA Tel: 0752 221312	BSc. Hons in Electrical & Electronic Engineering	3 years	
Polytechnic of Central London 18/22 Riding House Street LONDON W1P 7PT Tel: 01 486 5811	(a) BA Hons in Film and Photographic Arts	3 years	Theories of Art and Communication. Theory and criticism. Photography, film practice including technical process.

Address	Course title	Length of course	Subject area
	(b) BA Hons in Media Studies	3 years	Theory and practice of mass communication in journalism, radio, and TV.
Ravensbourne College of Design and Communication, School of TV & Broadcasting Wharton Road BROMLEY BR1 3LE Tel: 01 464 3090	(a) BTEC Higher National Diploma in TV Studio Systems Engineering	2 years	A broad-based course in the technology and engineering skills used when working as a Technician Engineer in TV Broadcasting, Radio Broadcasting, Closed-circuit TV, Satellite and Cable TV.
	(b) BTEC Higher National Diploma in Television Programme Operations	2 years	The production areas of cameras, sound, lighting, video tape recording and editing, vision mixing, telecine, audio recording and associated commercial practice.

Address	Course title	Length of course	Subject area
Sheffield Polytechnic Faculty of Cultural Studies Psalter Lane SHEFFIELD S11 8UZ Tel: 0742 556101	BA Hons in Fine Arts with Communication Arts option	3 years	Specialisation in 2nd and 3rd years in Film, Photography, TV for video and small studios.
Trinity & All Saints Colleges Brownberrie Lane Horsforth LEEDS LS18 5HD Tel: 0532 584341	BA in Public Media	3 years	A communication course with Film and TV options.
University of London Goldsmiths College Lewisham Way New Cross LONDON SE14 6NW Tel: 01 692 7171	Joint subject degree in Communications Studies with Sociology or Anthropology or single subject degree in Communications (from 1990)	3 years	Theoretical basis of communications, plus practice in Film, TV, Radio, Journalism, Electronic Graphics or Photography (students opt to specialise in one of the above).

Address	Course title	Length of course	Subject area
University of Surrey GUILDFORD GU2 5XH Tel: 0483 571281	B. Mus. Tonmeister	4 years	Music, Physics of Sound, Electronics, Electroacoustics, Recording Techniques.
West Surrey College of Art and Design Falkner Road The Hart FARNHAM GU9 7DS Tel: 0252 722441	BA Hons in Photography, Film and Video Animation	3 years	Photography, Film and Video Animation.
The University of Ulster Coleraine LONDONDERRY BT52 1SA Tel: 0265 44141	BA Hons in Media Studies	3 years	Film, Television (including Video Animation) and Radio, with one-third of the course practically based.

Courses for Postgraduates or Equivalent

Address	Course title	Length of course	Subject area
Bournemouth & Poole College of Art & Design Wallisdown POOLE BH12 5HH Tel: 0202 533011	Advanced Diploma in Media Production Options in (a) Film and TV (b) Multi-Media	1 year	Practical, self-exploratory course for people with a degree or BTEC HND in Photography or Film and Television.
Duncan of Jordanstone College of Art 13 Perth Road DUNDEE DD1 4HT Tel: 0382 23261	Postgraduate Diploma in Electronic Imaging	1 year	Application of micro-electronic techniques to manipulative images.
Falmouth School of Art & Design Wood Lane FALMOUTH TR11 4RA Tel: 0326 211077	Postgraduate Diploma in Radio Journalism	1 year	Radio Journalism, JACTRJ and CNAA (Council for National Academic Awards) Approved (see section on 'Other Employment Opportunities').

Address	Course title	Length of course	Subject area
Highbury College of Technology Dovercourt Road Cosham PORTS- MOUTH PO6 2SA Tel: 0705 383131	JACTRJ Diploma in Broadcast Journalism (exceptional non-graduates with relevant industrial experience may be considered)	1 year	News gathering, radio & TV reporting, editing, broadcasting, media law, etc. JACTRJ approved.
Lancashire Polytechnic PRESTON PR1 2TQ Tel: 0772 22141	Postgraduate Diploma in Radio and TV Journalism	1 year	News gathering, reporting, editing, broadcasting, etc. JACTRJ approved (see section on 'Other Employment Oppor- tunities').
London College of Printing Elephant and Castle LONDON SE1 6SB Tel: 01 735 8484	(a) Postgrad- uate Diploma in Radio and Television Journalism, Newspaper Journalism	1 year	Writing for radio, radio speech techniques, interviewing and reporting methods, the use of broadcast equipment, etc. JACTRJ approved (see section on 'Other Employment

Address	Course title	Length of course	Subject area
			Opportunities').
	(b) CNAA PG Diploma and Script-writing for Film and TV	2 years part-time	Research and Scriptwriting.
Middlesex Polytechnic Cat Hill BARNET EN4 8HT Tel: 01 368 1299	Postgraduate Diploma in Video	1 year	Video Production.
National Film & Television School Beaconsfield Studios Station Road BEACONSFIELD HP9 1LG Tel: 049 46 71234	Associateship of National Film and Television School	3 years	All aspects of Film Production, including writing, direction, producing, camera-operating, editing, lighting, sound.
Polytechnic of Central London 18/22 Riding House Street LONDON W1P 7PT Tel: 01 486 5811	(a) Postgraduate Diploma in Film Studies	2 years part-time	Theory and Criticism of Film.
	(b) MA in Film Studies	1 year part-time	

Address	Course title	Length of course	Subject area
Royal College of Art Kensington Gore LONDON SW7 2EU Tel: 01 584 5020	MA in Film and TV Production	3 years	Film and TV production.
University of Bristol 29 Park Row BRISTOL BS1 5LT Tel: 0272 303 204	Postgraduate Certificate in Radio, Film and Television	1 year	Intensive course in Film and TV, plus production of Audio and Videotape.
Centre for Journalism Studies University of Wales College of Cardiff 69 Park Place CARDIFF CF1 3AS Tel: 0222 874786	Postgraduate Diploma in Journalism Studies	1 year	Practical Journalism course includes broadcast specialism. Regular production days in Centre for Journalism Studies' own radio and TV studios.

Full details of these courses can be obtained from the addresses given.

Private Sector Courses

All the courses listed above are in the public sector. This does not mean that there are no worthwhile courses which are run by private organisations, but it is wise to exercise caution when choosing any course. Some of the questions worth asking are:

1 *Is the course vocational, academic or both?* A vocational course should give you practical skills which will enable you to do a job in the television industry such as editing or sound. An academic course should develop your appreciation of television as a medium of communication through discussion and analysis. Each has its role to play, but it is important that students should understand from the outset what the course is offering. A vocational course may help you to find your first job more easily (although you will need to be trained after appointment in the company's ways of operation). An academic course may help you in the longer term to develop your career in the creative side of the industry, but you will require more practical training before or after appointment to a job.

2 *Who are the tutors on the course?* Do they have professional experience in the industry? This is particularly relevant for vocational courses. It is worth remembering, however, that experience in the industry does not necessarily mean that the tutors were successful at their work, or that they are good at passing on their knowledge and skill to others. A 'big name' guest tutor who visits for two hours is no substitute for a first-class course tutor.

3 *What equipment is available?* Again, this is particularly relevant to vocational courses. While it is not essential to learn on 'state of the art' equipment, most of the facilities should be of broadcast standard if the course claims to equip you to work in the broadcast television industry.

4 *How much time does each student spend on the equipment?* Some training courses do not have their own equipment, but book time on a particular piece of machinery at a facility house. This is expensive, and may leave little time for all students to gain sufficient 'hands-on' experience.

5 *What technical support is available?* Is there engineering support on hand if the equipment fails, or are you likely to waste valuable time waiting for it to be repaired?

6 *How much does the course cost?* Does it offer value for money? Television equipment is very expensive, and any course which

offers hands-on experience will also be very expensive, unless it is run at a college or university. Courses in the private sector can cost several thousand pounds and there is therefore all the more reason to check that you are getting value for money. Speak to people who have attended the course if you can, and check with the Personnel or Training Department at the local television stations to see if they know of the course.

For details of other courses refer to:

- *Directions* – a guide to practical short courses in film and video.
- *Film and Television Training* – undergraduate and postgraduate courses with a practical emphasis.
- *Studying Film and Television* – courses in higher education which include film and/or television as part of their syllabus. The emphasis is on theory rather than practice.

All are available from:

The British Film Institute Publications Department
21 St Stephens Street
LONDON W1P 1PL
Tel: 01 255 1444

(*Note:* The courses mentioned in these three publications are not necessarily known to the ITV companies, and no guarantee can be given of their quality.)

Training in ITV

Recruitment of Trainees

The ITV companies recruit a significant proportion of trainees to fill vacancies, in addition to experienced applicants. The department which has the vacancy will normally consider the balance of experience among the remaining staff and, where possible, a trainee will be sought.

There are certain departments in which trainee vacancies are more likely to occur. These include cameras, sound, editing and engineering. This does not mean, however, that the number of vacancies is high. It is most unlikely that trainees would be recruited from outside the company into posts such as Trainee Director, Trainee Floor

Manager, Trainee Vision Mixer or Trainee Production Assistant. Most of these posts require experience within the television industry, and competition among internal applicants is intense.

Trainee Schemes

A number of ITV companies run formal trainee schemes from time to time. The decision on whether such schemes should be run is made according to the likelihood of suitable vacancies occurring at the end of the training period. They may not, therefore, be run every year.

The number of trainees recruited into formal trainee schemes is usually small – between two and ten at a time – and schemes may be offered in areas such as journalism, production, engineering and management. They normally begin in September and applications for journalism trainee schemes in particular are considered at the beginning of the year. Journalism, management and some engineering trainee schemes may be for graduates only. Training schemes may be advertised in the local or national press, or companies may approach relevant colleges and universities for suitable applicants.

It is regrettably not possible to give firm information here on where or when such schemes may be run.

Individual Trainees

The majority of trainees who join the ITV companies do so as individuals rather than as part of a group undertaking a formal trainee scheme. Vacancies may occur at any time of the year in any company, and training is conducted mainly on the job.

Graduate Trainees

There are few posts which are open only to graduates. Some graduates, however, are employed in most jobs in ITV. Applicants of graduate calibre may be preferred for certain posts, for example in sales and marketing, design, journalism, management and engineering. Some companies may occasionally offer formal training schemes for graduates, as mentioned above.

Training Prior to Appointment

ITV does not run training courses for members of the public. If you feel that a training course would enhance your prospects of obtaining a job, you should refer to the relevant Typical Recruitment Profile in the section 'Job and Entry Requirements' for advice on which courses, if any, might be appropriate. You should also cross-refer to 'Pre-Entry Education and Training Courses'. The majority of these courses are offered at colleges and universities in the public sector. Sponsorship is not normally available.

Training After Appointment

If you are successful in obtaining a post within ITV, you will be provided with the necessary training after appointment. This is likely to be structured on-the-job training, combined with attendance on short courses where appropriate. Further training will be provided as necessary throughout your career.

Employment

Trainees are recruited in anticipation that employment will be available in the company on completion of the training period. It is not possible, however, to give a complete guarantee of employment.

Training Courses for Employees

A wide variety of training courses is available to the employees of ITV companies. These courses may provide them with the basic skills and knowledge needed to do the job, to develop into more senior posts, to cope with changes in the technology or to adapt to the re-organisation of duties.

Many of these courses are organised by individual ITV companies for their own staff according to current needs.

Where there is a common need across the network, the ITV Association organises courses to which any of the ITV companies may subscribe. The Association currently offers approximately 60 courses a year ranging from basic training for journalists and programme directors to development skills for women, stereo sound training and management and supervisory skills.

ITV also uses various external organisations to provide training courses for its staff including Ravensbourne College of Design and Communication and the National Film and Television School.

(*Note:* These courses are open to ITV employees who are sponsored by their companies. They are not open to members of the public.)

WORK EXPERIENCE

Many courses of study include a period of work experience as a compulsory part of the programme.

ITV recognises the importance of such experience both to the individual and the industry, and each year several hundred students are given attachments with the ITV companies.

Regrettably, the demand for work experience far outstrips the supply of suitable opportunities for training, and ITV must therefore restrict attachments to those applicants who will gain the most from them. The criteria which are normally applied are as follows:

1 Applicants must be students following a recognised course of study (such as a degree or BTEC), at a college or university.

and

2 The course which they are following must be vocational, leading to the possibility of employment within the television industry. Ideally, work experience should be a compulsory part of the course.

and

3 The student must be resident in the transmission area of the company offering the attachment or, in some cases, attending a course in that area.

ITV is not normally able to offer work experience to students who are studying subjects which are unrelated to the work undertaken by a television company. Opportunities may, however, occasionally exist for students following computing, librarianship, finance, legal, administrative or management courses.

Applications from teachers and lecturers for attachments to their local ITV station may be considered if the courses for which they are responsible are particularly relevant to the television industry.

Applications from school children are not normally accepted owing to the heavy demand for attachments from students, and the lack of opportunities to provide suitable experience.

Work experience attachments may vary in length from half a day to several weeks or months depending upon individual needs and the availability of suitable opportunities for gaining experience.

Students do not normally receive payment from the company but students from certain designated courses which are of particular relevance to the industry are paid expenses. Students from sandwich courses who are on long-term attachments may be regarded as short-term employees and paid accordingly.

If you wish to apply for work experience and meet the necessary criteria, you should apply in writing to the Personnel Department at your local ITV company. Vacancies are not normally advertised and are rarely known far in advance. ITV regrets that it is never possible to guarantee an attachment, even if the necessary criteria are met.

BOOKLIST

General

Working in Television by Jan Leeming. Batsford Academic & Educational Limited

Film & TV: The Way In. A Guide to Careers. British Film Institute

Training and Careers in Film and Television. The Association of Independent Producers

Education and Training for Film and Television (3rd edition). Available from BKSTS, 110–112 Victoria House, Vernon Place, LONDON WC1B 4DJ.

Production Techniques

The Production Assistant in TV and Video by Avril Rowlands. Butterworths/Focal Press

Script Continuity and the Production Secretary by Avril Rowlands. Butterworths/Focal Press

The Small Television Studio by Alan Bermingham et al. Butterworths/Focal Press

The Technique of Lighting for Television and Motion Pictures by Gerald Millerson. Butterworths/Focal Press

The Technique of Television Production by Gerald Millerson. Butterworths/Focal Press

On Camera by Harris Watts. BBC Publications

Video Active by Geoff Elliott. BBC Publications

Television Sound Operations by Glyn Alkin. Butterworths/Focal Press

Using Video Tape by J. F. Robinson and P. H. Beards. Butterworths/Focal Press

Modern Recording Techniques by Robert Runstein. Howard Sams (UK Distributor, Pitman Publications)

Techniques of 3-Dimensional Make-Up by Lee Bagyan. Watson & Guptill (UK Distributor, Phaidon Press)

Stage Make-Up by Richard Corsan. Prentice-Hall (UK Distributor, IBD Publications)

Television Graphics by Douglas Merritt. Trefoil Design Library

Journalism

The Techniques of Television News by Ivor Yorke. Butterworths/Focal Press

News – Whose Bias? by Martin Harrison. Policy Journals

News, Newspapers & Television by A. Hetherington. Macmillan

Engineering

Colour Television Theory by Hutson. McGraw–Hill

Writing

Screenplay. Foundations of screenwriting. A step by step guide from concept to finished script, by Syd Field. Delacarte Press, New York (available from Ian Mead Ltd, UK)

Scripting for Video and Audio Visual Media by Dwight V. Swain. Butterworths/Focal Press

Writing for the BBC. BBC Publications, 35 Marylebone High Street,
    LONDON W1M 4AA
The Writers' and Artists' Yearbook. A & C Black (annual)

Note

There are many trade magazines which will give you a considerable
amount of information about television. Examples are *Broadcast,
Marketing Week, Television Week, Campaign* and *Audio Visual*.

The national press also carries many informative articles about
developments in television, as well as programme reviews.

Places to Visit

- Granada Studios Tour, Water Street, Manchester.
- The Museum of the Moving Image, South Bank, London.
- The National Museum of Photography, Film and Television,
  Prince's View, Bradford.

Many television companies will organise group tours of their
studios by special arrangement with their Public Relations Department. It is also possible to obtain tickets for audience shows from the
Public Relations Department of your local ITV company.

HOW A DRAMA PROGRAMME IS PLANNED AND MADE

The diagram opposite shows the various stages in the making of a
television programme, in this case a drama. Broadly similar proce-
dures are followed for all other types of programme. Note the
emphasis on planning and preparation, and the interrelationships
between the various jobs and departments.

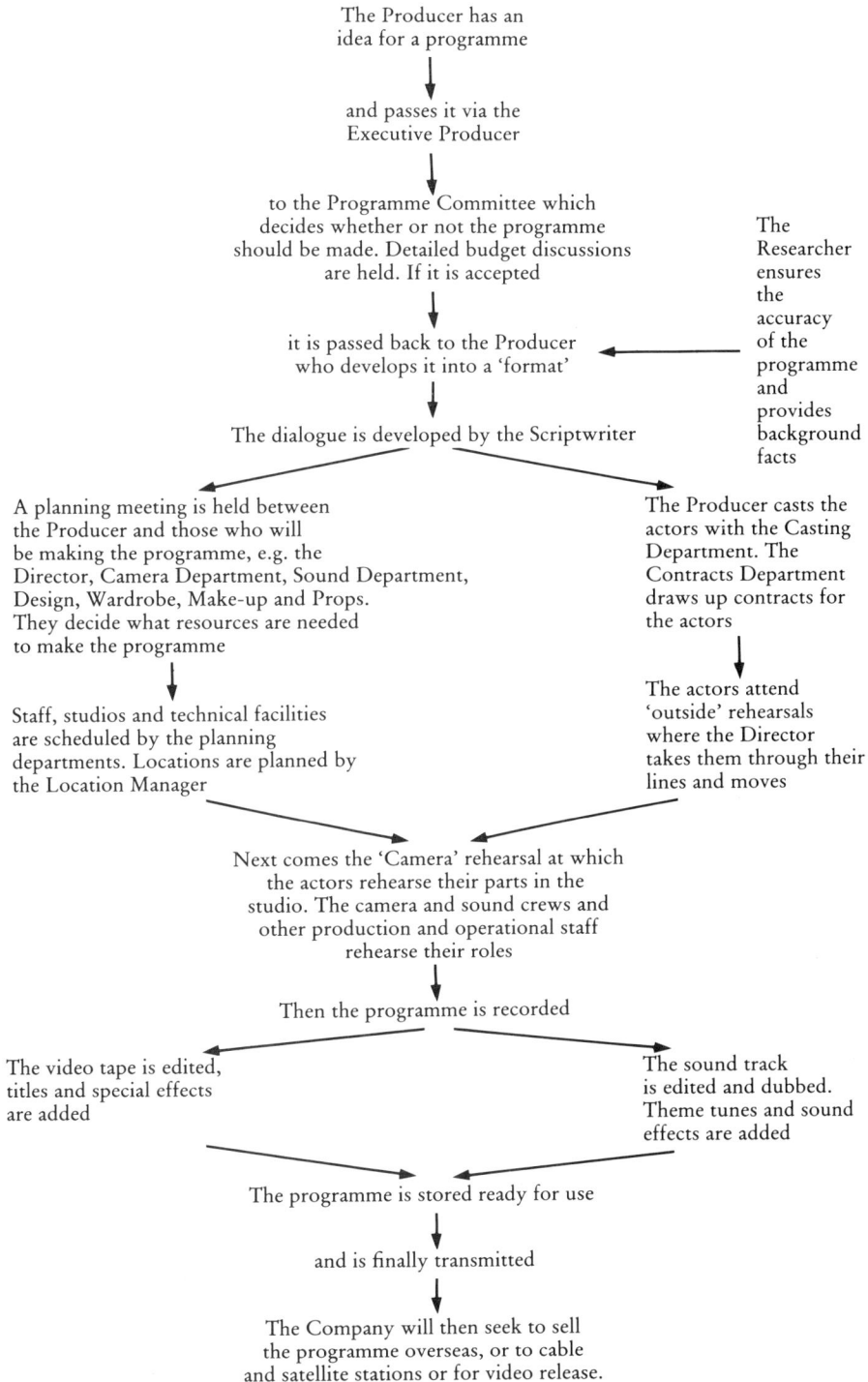

The Producer has an
idea for a programme

↓

and passes it via the
Executive Producer

↓

to the Programme Committee which
decides whether or not the programme
should be made. Detailed budget discussions
are held. If it is accepted

↓

it is passed back to the Producer
who develops it into a 'format' ←

The
Researcher
ensures
the
accuracy
of the
programme
and
provides
background
facts

↓

The dialogue is developed by the Scriptwriter

A planning meeting is held between
the Producer and those who will
be making the programme, e.g. the
Director, Camera Department, Sound Department,
Design, Wardrobe, Make-up and Props.
They decide what resources are needed
to make the programme

The Producer casts the
actors with the Casting
Department. The
Contracts Department
draws up contracts for
the actors

↓

↓

Staff, studios and technical facilities
are scheduled by the planning
departments. Locations are planned by
the Location Manager

The actors attend
'outside' rehearsals
where the Director
takes them through their
lines and moves

Next comes the 'Camera' rehearsal at which
the actors rehearse their parts in the
studio. The camera and sound crews and
other production and operational staff
rehearse their roles

↓

Then the programme is recorded

The video tape is edited,
titles and special effects
are added

The sound track
is edited and dubbed.
Theme tunes and sound
effects are added

The programme is stored ready for use

↓

and is finally transmitted

↓

The Company will then seek to sell
the programme overseas, or to cable
and satellite stations or for video release.

43

Part Two

JOB AND ENTRY REQUIREMENTS

ADMINISTRATION

When one thinks of the kinds of people employed in a television station, it is the programme makers who immediately come to mind. There are, however, many administrative staff employed by the ITV companies and it is undeniable that employment prospects are generally more likely to occur in administrative rather than programme-making areas.

The administrative functions of the television station provide an essential managerial and support role for the programme makers. They employ people at all levels from clerks to senior managers.

It is worth remembering that there are almost no opportunities in ITV for junior 'runabout' jobs which have traditionally been a way into, for example, the film industry.

Some of the main administrative areas, outlined below, are the Company Secretariat, Accounts, Purchasing, Employee Relations, Contracts, House Services and Transport, Education and Community Liaison, Public Relations, Production Planning and Scheduling, and Programme Planning.

The Company Secretariat

Most companies have a Secretariat which is responsible for the legal aspects of running the business. The Company Secretary is a senior manager who may well hold professional accountancy, legal or administrative qualifications.

The department usually employs a small number of staff who look after the needs of the Board of Directors by preparing minutes of meetings and writing reports and papers. They also ensure that statutory records are provided to various government bodies and that legal aspects of the purchase, sale and administration of property are correctly carried out. They also look after the insurance of the company's property and staff. Previous knowledge of television can be helpful, but is not essential.

Accounts

The Accounts function of a television station falls very broadly into

Office staff at work in the Accounts Department (YTV)

two sections. First, there is the section which deals with the finances of the company, frequently known as Financial Accounts. Staff in this area assist in planning and controlling departmental budgets, providing statutory and internal financial reports, controlling payrolls, etc. Senior members of this section are normally qualified Accountants but administrative staff with an aptitude for figures are also employed.

Secondly, there is the Programme Accounts section which is responsible for estimating and monitoring the cost of making programmes.

On the whole, the Accountants who are employed in this section have some previous experience in film or television since they may be asked to cost a script with very little assistance from the Director. A knowledge of how programmes are made becomes essential when the Programme Accountant is asked to estimate the likely cost of using a particular location, of buying or making particular props or comparing the cost of making the programme on film, in the studio, or on outside broadcast. In some companies at least, part of this role will be handled by a Production Manager, although that title may cover a variety of different tasks. Programme Accountants are usually qualified Cost Accountants.

Purchasing

This department is responsible for the purchase of office equipment and general materials for the smooth running of the company. Staff employed in this area are responsible for receiving purchase requisitions from departments, for ordering items from suppliers, and for receiving and checking invoices for payment. Industrial purchasing experience is useful for senior posts.

The Purchasing department is not normally concerned with the purchase of items for inclusion in programmes. This is the function of the Production Buyer.

Employee Relations

The Employee Relations Department consists of the Personnel, Recruitment, Training, Welfare, Safety and Industrial Relations functions. The title of the department may vary from company to company and certain functions may be grouped together. In general, however, the main duties of the department are very similar to those in any other organisation and include the provision of a recruitment service, the organisation of training, job grading and review, salary administration, the keeping of Personnel records, counselling and advisory services for staff, the implementation of safe working practices, and the provision of specialist advice to line managers on the maintenance of good industrial relations. In some cases the administration of the Pension Fund may also be carried out by the Employee Relations function

Managers in this area will normally have appropriate professional qualifications such as those awarded by the Institute of Personnel Management (IPM). It is not uncommon for administrative staff in the Employee Relations function to progress into management having obtained professional qualifications.

Contracts

Actors, dancers, musicians and most other people who appear on television are not permanent employees of the ITV companies, but are contracted on a temporary basis for the duration of a programme or series. It is the responsibility of the Contracts Department to organise the payment of their wages and other conditions of their employment. Many programmes are repeated and the Contracts Department then arranges for 'residual rights' payments to be made to the various artists who appear. In addition, payments have to be made to musicians whose recorded work has been chosen for use in a programme, perhaps as a theme tune or as incidental background music. Payments are also made to authors whose work is used, for example, as a basis for a drama series.

Many freelance Directors, Producers, Camera Operators, etc, are employed by ITV. They must also be issued with contracts and paid accordingly.

In some companies the Contracts Department may be responsible for casting for programmes, although in larger companies, this is generally the responsibility of the Casting Director.

A thorough knowledge of contract law, copyright law, employment law and the negotiated agreements between ITV and Equity, the Musicians' Union, the Writers' Guild, or the various staff unions is essential. Considerable experience in the television industry is required for posts in this department, and/or a legal background.

House Services and Transport

The House Services Department is responsible for the provision of office and other general facilities. Tasks may include the allocation of office space, the planning of future expansion, the organisation of telephone systems and cleaning services, the purchase of office furnishings and the provision of postal and printing services. Jobs may be filled internally or externally and previous experience in television is not essential.

Security Officers may also be employed by the House Services Department, although in some companies they are employed on contract. Security staff are responsible for the protection of all the company's premises and equipment, and may typically be former police officers or former employees of major security companies.

The Transport Department is responsible for the smooth operation of the company's vehicles including film crew cars, outside broadcast scanners, and an assortment of other vans, lorries and cars. Senior posts are normally filled by applicants with experience in the buying and selling of vehicles in bulk, and an up-to-date knowledge of transport law.

Education and Community Liaison

A small number of people, often with previous teaching experience, are employed to liaise with schools, colleges and local authorities on the use of ITV programmes for education purposes. They may also carry out general liaison duties with the viewing community, usually in conjunction with the Public Relations Department. In some companies, this department is closely associated with the making of educational programmes and there may be opportunities for Education Officers to be Producers on such programmes. In other companies, the department is more closely aligned to the Public Relations function. Some previous contact with the television industry is desirable.

Public Relations

The Public Relations Department is responsible for all aspects of the presentation of the company to the public. This may include such things as the preparation of publicity material, liaison with the community, the promotion of programmes and the issuing of press statements. There seems little doubt that Public Relations is one of the growth areas in television at present.

A background in journalism can be useful, or previous experience in public relations or advertising. Secretarial or clerical posts in the department may provide career openings, but may need to be coupled with further education and professional training.

Production Planning and Scheduling

The Production Planning Department is responsible for the long-range forecasting of the requirement for studio time, equipment and staff in order to meet the company's commitment to make programmes. The Production Scheduling or Production Control Department is responsible for allocating studio time, equipment and staff to each programme and for monitoring progress until the programme is completed.

There is a variety of administrative and managerial jobs in these areas, some of them requiring more experience in television than others. All, however, involve considerable contact with other departments in the company and staff will be expected to develop a thorough knowledge of staff trade union agreements in operation in ITV.

Programme Planning

This department is responsible for planning the schedule of programmes to be shown in the local region. There is a considerable amount of liaison with other ITV companies to coordinate the transmission of networked programmes, with the Sales Department for the scheduling of advertising breaks, and with the Presentation Department for the scheduling of promotions, announcements, etc. Senior posts are normally filled internally, but more junior vacancies may be filled by external applicants with clerical, secretarial or administrative experience.

It is not possible to give here more than a general idea of the kinds of qualifications needed by applicants for administrative posts. More senior positions in the Legal, Personnel and Accounts departments, for example, may demand the appropriate professional qualification. Applicants for middle-graded posts will generally have a good record of GCE 'O' level or GCSE passes (grades A to C), including English and Mathematics, and ideally 'A' levels in any subject. Junior posts may be filled by applicants with 'O' level passes and GCSEs (grades A to C). Once again, English and Mathematics are most useful. All posts are likely to demand a high degree of accuracy, attention to detail, and the ability to develop and work to systems. The ability to communicate effectively in writing, on the telephone and face to face is also important.

See also: 'Librarians', Computers', 'Management', 'Secretaries and Clerks'.

ANNOUNCERS AND PRESENTERS

Announcers

The job of the television Announcer must seem to the public to be one of the most glamorous jobs available, and this is reflected in the overwhelming response received by the television companies whenever such jobs are advertised.

It is certainly true that the job can bring instant public recognition, with offers to make appearances at fetes and local shows, etc., and all the pleasures and problems that this entails.

There is, however, much more to the job than meets the eye. Announcers must be prepared to accept very considerable disruption to their private lives as they offer a service to the public which covers twenty-four hours a day, seven days a week.

The most obvious task undertaken by Announcers is the provision of a link between programmes, introducing the next programme to be screened, and previewing others to be shown later on. Within the industry, they are frequently known as Continuity Announcers, and this reflects their role of ensuring a smooth progression from one item to another.

The job involves a great deal of behind-the-scenes preparation and Announcers sometimes write their own scripts, although this may be done by Promotion Script Writers. They must therefore be able to write clearly and concisely for the spoken rather than the written word. They must also make sure that they are fully aware of any problems or events that might mean an interruption to normal services. A newsflash, for example, may have to be read in the middle of a programme, and there will be very little time to prepare a script. If there is a technical breakdown, the Announcer must keep the viewers informed and the ability to 'ad lib' can be very useful to fill a gap before normal services are restored.

Other duties may include on-screen interviewing, reading scripted commentaries for various programmes, doing 'voice-overs' on television commercials, reading news bulletins, compering programmes and carrying out research and preparing scripts for programmes.

Announcers also have a very important sales role to play, since the aim of all television companies is to reach the maximum number of viewers. The manner in which Announcers introduce or preview a programme can have a significant effect on whether viewers decide to watch. They must therefore have a broad range of interests and a good general knowledge so that they can introduce all types of programme from soap operas to documentaries with genuine interest and understanding.

Announcers, more than any other staff, represent to the public the image of the television companies for which they work. The programmes may come from any of the ITV companies or an outside production company, but the viewers know that the Announcer is part of their local ITV station. It follows that the different companies may look for very different kinds of Announcer to reflect their own image.

There are, however, certain qualities which all Announcers should possess. They must have poise and presence, and be able to speak with an air of authority. They must have a warm, friendly, approachable personality, an attractive appearance, and a clear well-modulated voice which is pleasant and easy to listen to. Regional accents are acceptable and, indeed, may help to reflect the local image of the company.

Experience has shown that the best Announcers have often received speech and drama training, and preference is given to applicants with experience in the entertainment industry. From time to time applicants without such experience are successful, and trainee vacancies are sometimes offered to people who are totally new to television announcing. The introduction of all-night programmes has increased the opportunities for younger 'character' Announcers.

Contracts of employment may be for staff posts, or they may be temporary for periods of several weeks or months.

Presenters

Television Presenters are responsible for taking the lead in, or 'fronting' programmes. They are the well-known personalities who are associated with certain news, sport, current affairs programmes, documentaries and quiz shows, etc.

It is of course impossible to describe the particular qualities required by Presenters, or a typical background from which they are recruited. Each will be different, according to the programme on which they are working.

Presenters on news and current affairs programmes will almost certainly have experience in journalism and many will have progressed from on-screen reporting on regional news magazine programmes. Reporters may indeed move on to become Presenters on a wide variety of programmes.

Presenters on quiz shows and other light entertainment programmes are usually experienced actors and actresses, while Sports Presenters are often well-known sporting personalities.

Documentaries are frequently presented by an expert on the subject under discussion, rather than someone with a television or theatrical background.

It follows that this is not a job for which you can apply in the usual way. Presenters are employed on a contract basis for the duration of the series, and they are normally approached by the company rather than the other way around.

No recruitment profile is given for this job.

Typical Recruitment Profile

Trainee Announcer

	Essential	Desirable
Physical		
Age	21 +	
Speech	Clear, well modulated	Speech training
Appearance	Smart and attractive Acceptable to all types of people	
Education	Minimum of GCSE Grade A to C English language	'A' levels or degree
Experience		Theatrical experience highly desirable Membership of Equity

	Essential	*Desirable*
Interests	A wide range of interests of all kinds. Knowledge of television programmes. Wide range of reading interests	
Personality	Warm, outgoing and friendly but authoritative personality. Poise. Able to remain calm and think quickly under pressure	
Availability	Able to work unsocial hours	

Note: Remember that it is how you look and sound to the camera that matters, not how you appear and are heard in the flesh. It may be worth investing in a short video of yourself showing what you can do.

CAMERA OPERATOR

The job of the Camera Operator undoubtedly presents to the outsider an impression of glamour and prestige. Both of these it certainly has to a degree, but as with most jobs in television, the work is demanding, with a long period of training. By no means all of a Camera Operator's time is spent on the camera. They have plenty of other duties, many of which are strenuous, sometimes dirty and often uncomfortable.

The majority of programmes seen on television are pre-recorded on video tape from pictures shot on electronic (video) cameras. Electronic cameras are also used for the relatively few 'live' programmes that are shown (mostly news, current affairs, and chat shows).

A comparatively small number of programmes, however, are shot on film. The main examples of film on television are some, though not all commercials, some full-length drama features, both old movies and those produced for television, and some documentary material. Film possesses certain qualities which have in the past been lacking in video tape; however, advances in technology have enhanced the quality of video. Since it is generally cheaper and more convenient to use, an increasing number of programmes are being made on tape at the expense of film. This should pose no real dilemma for a young person wishing to start a career in film camera operations, for it is relatively easy for the skills acquired in film to be transferred to electronic (video) cameras.

Nowadays, the word 'film' is used quite haphazardly to describe any television material that is not being transmitted live. One also hears the occasional reference to 'live filming', e.g. of the Cup Final. Such inaccuracy is pardonable since it conveys a concept that is easily understood. However, it is highly probable that the programme has been made on tape, not film.

For the sake of clarity, we shall describe here the work of the Camera Operator under four headings: Studio, Outside Broadcast (OB), News, and Film.

Studio

A typical studio may have three, four, five or even six electronic cameras in operation at one time, according to the kind of programme that is being made. Most of the cameras currently in operation are fairly large and are positioned on a pedestal, supported on mobile mountings. The mountings glide smoothly over the floor and enable the Camera Operator to position the camera for any kind of shot.

Each camera is operated by one person, but there are other supporting members of the team or 'crew'. Some studio cameras, for example, are mounted at the end of a jib on a motorised crane which can be raised to a considerable height, or lowered to within an inch or two of the floor.

In addition to the Camera Operator who is seated at the end of the jib operating the camera, there may be a second member of the team whose task is to 'swing' the counterbalanced arm. A third member, seated at the back of the crane, is responsible for driving it accurately to precise positions marked on the studio floor. The tasks of driving and swinging the crane are just as much a part of camera work as operating the camera.

Lightweight studio cameras can also be mounted at the end of a long crane arm, and operated by remote control by a Camera Operator on the studio floor – a different skill again.

The majority of cameras currently in use in studios have the encumbrance of long cables which snake across the floor to sockets in the studio, and via them to the control equipment. These cables have to be constantly man-handled to allow the cameras freedom of movement. This task is known as 'cable bashing'. In some companies, cable bashing is the responsibility of the junior members of the camera crew, while in others, everyone takes their turn.

An Electronic Camera Operator working in the studio (or on OBs) generally has less discretion than a Film Camera Operator in how the action will be shot. The Programme Director will normally decide the relative positions of the cameras, and the size of shots. During the camera rehearsals for a recorded programme, the Camera Operator will practise taking up these positions, following a check-list of agreed shots which is attached to the camera. During the recording, the Production Assistant will remind the Camera Operator via the 'talk-back' system of the shots which are coming up.

Studio camera operator (STV)

Although some very experienced Directors will dictate the precise composition of the shots, most will simply call for a 'close-up' or 'two-shot' or whatever and the framing and composition of the shot will be left to the Camera Operator. A skilled Camera Operator will anticipate the kind of shot that the Director is seeking and 'offer' the shot on the studio monitor for selection. This is particularly useful during live shows when there is little chance for rehearsal.

Another of the Camera Operator's skills is the ability to reposition the camera quickly and frequently without causing obstruction to the other cameras, microphone booms, and artists. Excellent hand–eye coordination is needed for zooming, etc.

The work can be very tiring as Camera Operators must stand for long periods under hot studio lights. They must also have the personal qualities needed to work as a member of a team. They must be able to help artists to give their best performance since the working relationship between the artist and the Camera Operator is often very close.

Smaller cameras are increasingly being used in the studio, and may be hand-held on the shoulder, or mounted on a pedestal as necessary. In certain cases, such as some news bulletins, the operation and positioning of these cameras may be remotely controlled by computer so that there is no Camera Operator on the studio floor.

Outside Broadcasts

The traditional 'OB' unit is a mobile studio control room with a number of cameras linked to it. Such units are mainly used for state occasions, sports fixtures, political conferences, and other events which take place away from the studio.

There are, however, an increasing number of 'mobile' units which make use of small, lightweight cameras to cover OBs, and to record inserts for dramas and other studio productions. These units may consist of several cameras, or just one portable single camera (PSC) which is used by the Camera Operator in much the same way as a film camera.

Whatever the kind of unit, there are common problems which the Camera Operator working outside the studio must face. These include finding a suitable place to position the camera, which may be on a scaffolding tower, a roof or a grandstand in less than ideal

Recording *World in Action* on a studio set of the House of Commons (Granada)

conditions. The weather can also be a problem and the Camera Operator must be prepared to work in the cold, the wind, and the rain.

News

News camera operations require a special kind of individual. They must not only be technicians, but both individualists and team players. They are leaders, artists and entrepreneurs who often live on their wits and may have to rely on other members of the news crew for their safety. The 'street-wise' skills of News Camera Operators are often as important as their technical and artistic abilities. News is often unpredictable, and Camera Operators cannot know in advance what difficulties and dangers they may be asked to face, or what human emotions they may have to confront. A high degree of fitness and stamina is essential.

There is no Director to decide what shots are to be taken for a news

An ENG crew recording the Piper Alpha disaster, July 1988 (ITN)

item. It is the responsibility of the Camera Operator to seek out the most telling pictures.

The title of Camera Operator is becoming something of a misnomer in relation to news. They may work in a team of only two, carrying out all the functions of cameras, sound, lighting, simple editing, first-line maintenance of their complex gear, manning electronic communications links equipment, and playing material along lines and satellite links back to base.

In some cases they may work entirely alone, carrying out as many of these functions as are practical at any given time with ENG (Electronic News Gathering) equipment. In either case, they must use their initiative since they have little of the support that is available in the studio.

Entry into the profession can be via a number of routes. In most regional ITV companies, news is combined with other camera operations such as studio, OB and film, and Camera Operators may be rostered onto news for several weeks at a time. At ITN, however, News Camera Operators are specialists. Some may have begun their careers as studio, OB or film Camera Operators, and some may have gained experience in sound or lighting. Some are appointed directly from outside if they are able to demonstrate that they are 'naturals' for on-the-road work.

Most ITV companies use the services of 'stringers' from time to time. These are freelance News Camera Operators who cover a particular geographical region and can be quickly on the spot when a story breaks. They may send pictures to more than one company or news organisation.

▌ *Film*

At one time, film was used extensively on television, and Camera Operators were employed exclusively on film. This is rare today, and most Film Camera Operators also use lightweight video cameras in their work. There are, though, still a small number who specialise exclusively in film.

The top jobs in film cameras are those of the Lighting Camera Operators. They are the senior technicians on film units and are responsible for both the technical and artistic quality of the pictures. With the Director, they make the major decisions about the positioning of cameras, the way in which the action is shot, and how the scene will be lit to meet the needs of the action. In this respect, Film Lighting Camera Operators tend to have more artistic discretion than Electronic Camera Operators. Lighting Camera Operators also supervise the work of the camera crew. On a big 'shoot' where several scenes or locations have to be filmed in sequence, they may well leave the operation of the camera to one of the crew while doing the lighting for the next action-area. Lighting Camera Operators tend to be specialists in either documentaries or dramas and when working on an assignment, will normally be away from base for several weeks, travelling extensively and coping with novel situations.

Exterior Camera Operators are the next rung down the ladder. They do much the same job as Lighting Camera Operators but usually on less prestigious productions, perhaps specialising in documentary work. In some companies, however, the grade of Exterior Camera Operator no longer exists.

Next are the Camera Operators who, normally under the supervision of a Lighting Camera Operator or Lighting Director, work the camera on every kind of shot.

The Camera Assistants, or 'Focus Pullers' or 'Clapper Loaders', are the least experienced of the crew. They are responsible for the care and cleanliness of the camera, loading and unloading film, and keeping records of all 'takes'. On long tracking shots where the camera has to move towards or away from the action, the Camera Assistant operates the focus control on the camera. After starting as a trainee and working for nine months under the supervision of another Assistant, plus about another two years alone, the Camera Assistant will 'act up' as Camera Operator, coping with ever more complex shots.

Working on a film unit involves a good deal of travel and long periods spent away from home on locations, often in bad weather. It is not the job for anyone who likes an ordered existence or a comfortable life.

Typical Recruitment Profile

Trainee Camera Operator

	Essential	Desirable
Physical		
Age	18 +	20 +
Health	Physical fitness Stamina and agility	
Eyesight	Good colour vision	
Hearing	Good (able to hear instruction clearly through headphones)	
Coordination	Excellent coordination of sight, touch, hearing and general movement	
Speech	Clear	
Education and Training	Broad general education to at least GCSE standard (grades A to C)	GCSE grades A to C in English, Maths, Physics. 'A' levels (subjects not specified). Knowledge of lighting, optics. BTEC HND in Programme

	Essential	*Desirable*
		Operations. Film and television or photographic course at college of further education. An understanding of electronics is useful for work away from the studio
Interests	Practical evidence of interest in and flair for photography (preferable cine or video), from artistic and technical point of view	Amateur dramatics, current affairs, TV, films, theatre
Personality	Equable, sociable temperament. Able to work as a member of a team, and act on own initiative	
Availability	Able to work irregular hours at any location when required. Able to work away from base for long periods when required	

Note: 'Essential' qualifications are rarely sufficient for entry into this job. Most successful applicants have 'Desirable' qualifications.

COMPUTERS

All of the ITV companies use computers, and most have their own computer departments. The titles of these departments vary, but they may be described as Computer Services, Management Information Services, or Information Technology Services for example. Most such departments are small, and employ no more than a handful of people. Some are continuing to grow as new computer systems are introduced.

Large computers (mainframes) have been in use in ITV for many years, often linking the various companies. Others link the Sales Departments (which sell advertising time) with leading advertising agencies.

Medium-sized computers (minicomputers) and small personal computers or PCs have been increasingly used in recent years. They frequently link individuals and departments within the company.

The applications of computers in television are too numerous to mention, but they include

- Accounting and financial management
- Programme planning and production planning
- Controlling the sales of advertising time
- Personnel records
- The preparation of programme scripts
- Graphic and set design
- Designing new equipment
- Controlling automatic studio equipment such as lighting
- Library records and control
- Word processing
- Newsroom operations

Most engineers employed in ITV need to be able to use computers, and to carry out computer programming in engineering computer languages. As new technology advances, the engineers and the specialists from computer departments increasingly need to work together. Computer specialists must keep up to date with the new technology, and must of course learn how a television station operates. They have regular contact with people in all departments where computers are used, and must be able to communicate technical concepts clearly and effectively to non-technical people.

Opportunities for employment may arise at a number of levels. Business Analysts and Systems Analysts may be employed by some companies to investigate organisational and management information problems, and to decide whether computers can offer a solution.

Programmers may be employed to write instructions for the computer to enable it to perform the desired task, while Computer Operators will load and operate the machinery. Specialist Operators may be employed on large systems to type in large quantities of information, and distribute printouts.

It is sometimes possible to join as a Trainee or as an Operator and to progress upwards to Programmer, Analyst, or Operations and Systems Manager level. This can, however, take a number of years and cannot usually be done without studying for professional qualifications. Higher Diploma or Degree qualifications are common at more senior levels, as are qualifications awarded by the British Computer Society.

Typical Recruitment Profile

Trainee Computer Operations Assistant

Note: This level of entry is not covered in the text but opportunities may arise occasionally for young persons not long out of school in companies that are large computer users. As computer applications for routine business increase, there may possibly be more opportunities for entry at 16+ with limited qualifications.

	Essential	*Desirable*
Physical		
Age	16+	
Physical	As for Trainee Operator	
Education and Training	GCSE grade A to C Maths & Computer Studies *Note*: (1) Some Government approved Youth Training Schemes, e.g. those run by ITEC Training Centres, may be	

	Essential	Desirable
	acceptable. (2) Some companies may accept trainees with lower academic qualifications who are able to pass an aptitude test in logical thinking	
Personality	As for Trainee Operator	

Typical Recruitment Profile

Trainee Computer Operator

	Essential	Desirable
Physical		
Age	18 +	
Eyesight	Good colour vision	
Hearing	Good. Able to take instructions in a busy room	
Coordination	Good hand–eye coordination. Manual dexterity. Capable of learning to touch-type	
Education and Training	Advanced Level GCE in Computer Studies	BTEC Higher Computer Studies. Knowledge of electronics or physics. GCSE grades A to C Physics, English, Maths
	Note: (1) Some Government funded long-term training schemes requiring 4–5 good GCE 'O' Levels (or GCSEs) may be a substitute.	

	Essential	Desirable
	(2) Some companies may accept trainees with lower academic qualifications who are able to pass an aptitude test in logical thinking	
Personality	Able to work as a team member. Calm temperament. Able to concentrate on detail	

Typical Recruitment Profile

Trainee Programmer/Analyst

	Essential	Desirable
Physical		
Age	20+	
Physical	As for Trainee Operators	
Education and Training	Advanced Level GCE Computer Studies Some companies may accept trainees with lower qualifications who are able to pass an aptitude test in logical thinking	Degree in Computer Science & Student Graduate of British Computer Society (Minimum entry two 'A' Levels)
Personality	As for Trainee Operators plus an analytical problem-solving mind	

Typical Recruitment Profile

Business/Systems Analyst and Operations Management

	Essential	*Desirable*
Physical		
Age	There is no definable entry age	
Physical	As for Trainee Operators	
Education and Training	Degree in Computer Sciences, or Full Membership of British Computer Society, or Combined Degree in Business Studies and Computing	Second degree or comparable qualifications in Business Studies if not incorporated in first degree. Four to five years relevant experience and some line management experience
Personality	Calm temperament. Inquisitive and analytical. Strong interpersonal skills; able to interview 'client colleagues'	
Interests	A wide range of business and technical activities	

COSTUME DESIGN AND WARDROBE

Independent Television has produced many memorable drama productions in which the magnificent costumes have contributed much

Wardrobe staff at work on location in Amsterdam (YTV)

to the impact of the programme. Costumes also play a vital part in many light entertainment shows, particularly those that feature dancers. Most of these programmes are, however, made by the larger ITV programme companies. Career opportunities in costume/ wardrobe in the smallest companies are limited. The demand for wardrobe facilities is so variable that even the major companies rely heavily on freelance and contract staff when they are making a major programme. Staff vacancies do, however, occur from time to time, particularly at the lower levels of the career ladder, and employees who can demonstrate talent and perseverance can progress to the top. There are a number of differences in the job titles used in the various ITV companies and in the content of jobs. For the sake of simplicity, we have used the title of 'Costume Designer', for example, but in many companies similar tasks are carried out by the Senior Wardrobe Supervisor.

The most senior posts are held by Costume Designers. They usually work on the most prestigious programmes and their responsi-

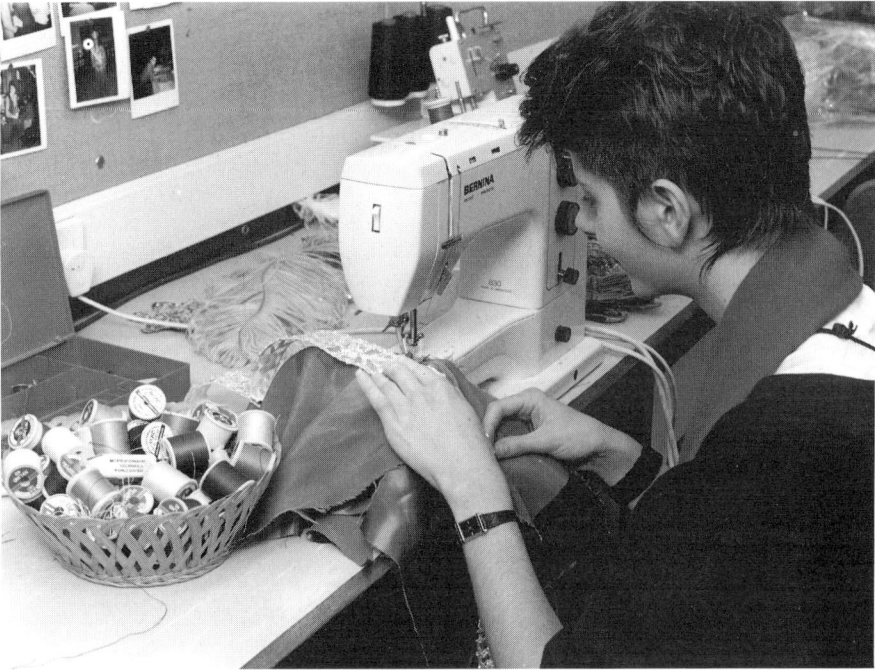

Working in the Wardrobe Department (YTV)

bilities include planning, management of staff, designing, fitting and budgeting.

The planning stage is a vital one, and Costume Designers (or their equivalent) must read the script thoroughly before deciding what type of costumes will be appropriate. The wishes of the Programme Director must of course be taken into account and consultation must take place with the Set Designer to ensure that costumes and sets blend together in perfect harmony. The plans may have to be changed once the casting has been done as the chosen costumes may not suit the artist.

Costume Designers must plan the entire operation down to the last detail, including when changes of costume are going to be necessary. They may design costumes which are to be made up from scratch, or plan alterations to transform existing costumes or accessories.

It is of course impractical for television companies to carry large stocks of clothes and most costumes are hired from theatrical costumiers. Costume Designers visit the costumiers with the artists to select and fit suitable clothes. They may also supply the accessories and trimmings that will help to transform the clothes to reflect the right period or the right character. Any alterations that are needed to the clothes at this stage are usually done by the staff at the costumiers.

The job clearly calls for a high degree of creativity combined with administrative and supervisory skills, the ability to interpret written ideas, and the ability to work well with artists. They must also be prepared to spend long periods researching to ensure that costumes are authentic. They must possess a thorough knowledge of fabrics and costume styles appropriate to various periods. A thorough understanding of dress-making techniques is also essential. If the production is very large, they may be assisted by a Wardrobe Supervisor or an Assistant who will help with the fittings and the general running about. These staff carry out many of the same duties as the Costume Designer, except designing, and are responsible for wardrobe continuity. They often work on shows which require modern dress and are therefore frequently called upon to take artists shopping for clothes.

Below the Wardrobe Supervisor is the Wardrobe Assistant/ Dresser. One of the main functions of the Dresser is to help artists into and out of their costumes during the production and to have ready any changes of clothes that may be necessary. This part of the job calls for tact and discretion as many artists are under considerable stress before a performance. Dressers must have the ability to sense how artists wish to be treated. Some will enjoy a chat while they are being dressed, while others will prefer a courteous silence. It is important that Dressers are confident, but not over-confident, and not too familiar.

This same sensitivity to other people is of course an essential qualification for every grade in the Costume and Wardrobe Department.

Dressers also carry out minor alterations to costumes, sometimes at the very last minute, and the ability to sew quickly and neatly is therefore important. They are also responsible for the general care of clothes and accessories including washing, cleaning and pressing.

Since Dressers do not design clothes or carry out major alterations, they do not require tailoring or design training. If they wish to move up the promotion ladder, such qualifications will become increasingly important.

All grades of staff may be called upon to work in the studio and on location. Unsocial hours are common, and location work frequently means working under difficult conditions.

Typical Recruitment Profile
Wardrobe Dresser

	Essential	Desirable
Physical		
Age	18 +	22 +
Health	Good	
Vision	Normal colour vision	
Coordination	Hand to eye	
Appearance	Clean, neat and tidy. Good personal hygiene	
Education		GCSEs (grades A to C) in English, History, Art, Mathematics, Dressmaking.
Experience	Employment in theatre or film Wardrobe, or with a theatrical costumier	
Interests	Creative fashion. Art, television, theatre, history	
Personality	Tactful and discreet. Sensitive to the needs of others. Able to work under pressure. Sociable, practical	
Availability	Irregular hours, some work away from base	

Note: Applicants who wish to progress to more senior grades should preferably have attended an Art College course in fashion or theatre design, or an apprenticeship in tailoring. Creativity and artistic ability is essential for senior posts together with an excellent sense of fashion.

CRAFTS AND TRADES

There are a number of crafts and trades within Independent Television. Some of the staff are employed solely on site maintenance and some are specialist craft workers who apply their skills to television production. Those employed on site maintenance cover the normal range of duties associated with their trade.

Not all ITV companies employ staff in these grades. Some contract out the work.

Carpenters and Joiners

Carpenters are employed in the general installation and maintenance of all wooden items in offices and other premises.

Others are employed in the construction of sets for television programmes and may be called upon to make an endless variety of items from antique furniture to the interior of a space ship. Most of the sets are dismantled after use, so there is little point in using all the skills of the craftsman to make them last. Some items, such as chairs, however, must be able to stand a fair amount of wear and tear and must be made to the highest standards. Carpenters work from drawings made by the Set Designer and the ability to interpret drawings accurately is most important, as it is in all crafts.

Painters

Painters are employed for the painting and decorating of offices and other premises.

Others are employed to paint the sets for television programmes and they are called upon to use a variety of skills not normally required of a painter and decorator. They may paint 'walls', 'doors' and 'floors', etc., under the guidance of the Set Designer. There is scope for considerable creativity when, for example, they are asked to turn a plain piece of wood into an 'antique door', or reproduce an elaborate 'plaster moulding' that will withstand the close scrutiny of the camera.

Drapes/Upholstery

Some of the larger companies employ upholsterers to make soft furnishings for studio sets. They follow detailed plans prepared by the Set Designer to make such things as curtains, or padded upholstery for the interior of a 'pub'. Only small numbers of staff are employed in this field.

Plumbers

Plumbers are of course employed to carry out the normal range of plumbing duties on the company premises and may maintain heating and ventilating systems, etc. They are also required to provide plumbing facilities for studio sets. They may, for example, be responsible for supplying water to the taps in the sink on a set depicting a kitchen.

Fitters/Mechanical Maintenance Engineers

These staff are responsible for the maintenance of any mechanical equipment on the premises including camera dollies, microphone booms, film editing machines, etc.

Mechanical Maintenance (YTV)

▌▌ *Recruitment*

Staff may be recruited into these posts as qualified craftsmen and tradesmen (these terms apply to both men and women) or, from time to time, apprenticeships may be available. Companies prefer to recruit apprentices locally and vacancies are normally advertised through local careers offices and local newspapers.

The apprenticeships last for three or four years and combine periods at college on day or block release with practical on-the-job training. Applicants are expected to have a good standard of education and entry qualifications are usually related to the further education course to be followed by the apprentice. GCSE grades A to C in English and Mathematics are both useful.

Although applicants for apprenticeships are expected to be 16 or 17 years of age, favourable consideration will be given to older applicants who have been made redundant from apprenticeships in similar fields in other industries.

No recruitment profile is given for these posts, but applicants for all jobs must be prepared to work irregular hours or shift work. They must also be prepared to travel away from base on occasions.

It is possible that in future, staff will cease to specialise in one craft or trade, but will become multi-skilled across a range of jobs.

Apprentice Carpenter working on a studio set (Anglia)

EDITOR: FILM AND VIDEO TAPE

The Editor's work begins when most of the rest of the production team have finished, and moved on to another production. The Editor's task is to assemble the constituent parts of the programme, remove the pieces that are not needed, and present the programme in its final form ready for transmission. The sound department may be required to carry out some work on the sound track after editing has been completed but in essence, it is the Editor who is responsible for creating the final form and style of the programme. Editing is therefore an essential function in all programmes, except those which are totally live or recorded continuously and require no 'post-production' work.

Film Editing

The fundamental difference between editing film and editing video tape is that film is physically cut, and video tape is not. Film usually requires a great deal of editing as programmes are frequently shot on one camera only. This leads to a considerable amount of out-of-sequence shooting. The types of programmes which are sometimes made on film are documentaries, and dramas made on location.

The Editor views the developed film frame by frame with the Programme Director and decides where cuts should be made. 'Takes' that were unsuccessful are removed, scenes are arranged in the correct order, and individual frames may be removed to create a better effect. The sound track is also cut to coincide with the picture. Once the cuts have been made, the film is re-assembled using transparent joining tape or a special cement to produce the 'cutting-copy'. The job clearly calls for a high degree of precision and artistic flair and meticulous attention to detail. Editors must be familiar with the technicalities of film processing and materials so that they can pass instructions to the film laboratories who will produce the final copy.

It takes a long time to build up the knowledge and experience needed to be a skilled Film Editor, and most begin their careers as Trainee Assistant Film Editors, with promotion to Assistant Film Editor after nine months.

Assistant Film Editors do not make editing decisions, but provide the support services needed by the Editor. Their main task is to

Film Editor (LWT)

synchronise the 'rushes'. The picture and sound tracks are separated and the original picture negative is stored away safely to keep it in prime condition until it is time for the final editing to be done. The Assistant Film Editor lines up a copy of the picture 'takes' with the sound tracks by coordinating the picture and sound of the clapper board. It is then ready for the Editor to work on. Other important duties are looking after the equipment in the cutting room, cutting and joining film, and keeping thorough records. There may be hundreds of strips of film in the cutting room, some waiting to be joined together and some waiting to be discarded. The Assistant Editor makes a note of what is on each strip, and the editing decisions that have been made. All discarded film is carefully logged and stored away so that it can be quickly located if the Editor suddenly decides to use it. Very neat handwriting, attention to detail, and an orderly mind are essential qualifications for the job. Promotion to Film Editor is not automatic, and will depend on ability and the availability of posts. It may take several years, but most experienced Assistant Film Editors will be given practical experience of minor editing tasks to help them to compete for promotion.

Video Tape Editing

The majority of television programmes are made in the studio where the output from several cameras and other visual sources is recorded directly onto video tape. The equipment used by a Film Editor is relatively simple, and hand operated. Video editing machines on the other hand are sophisticated pieces of electronic equipment and this explains why many present Video Tape Editors began their careers as television engineers and technicians. It is not essential, however, for a Video Tape Editor to have a technical background. The operation of the machine can be learned in a relatively short time and it is the ability to make the right editing decision that is important. The tape is contained on reels and once the Editor and Programme Director have decided where the edit points should be, the scenes to be retained are recorded onto another tape, the original remaining untouched.

Video Tape Editors may come from a variety of backgrounds but all will have some previous experience in television.

The role of the Assistant (or Junior VT Engineer/Editor) in video tape editing is similar to that of the Assistant Film Editor. The Assistant is responsible for 'lining-up' the machine, i.e., setting up the controls ready for use, and the general fetching and carrying of tapes, etc. It follows that a technical background is essential and entry is normally through operational engineering routes. Promotion to Video Tape Editor is possible for those with the necessary qualities. Not all companies employ staff in the Assistant grade.

Editing Single Video Camera Production – PSC Editing (ENG and EFP)

Single video camera production is often referred to as PSC (Portable Single Camera) for short. It may also be referred to as EFP (Electronic Field Production) or ENG (Electronic News Gathering), according to the type and scale of production. Single camera productions combine the use of video tape with techniques of shooting which are in many ways similar to shooting film. As in film, only one camera is normally used, so there is much out-of-sequence shooting. This has implications for the Editor.

Single camera operations for news are generally known as ENG. The technique of editing ENG (or PSC) is very similar to that of editing video tape, the main difference being that small cassettes are used rather than large reels of tape. The scenes to be retained are recorded onto a separate cassette and the original remains untouched.

Any Editor will work under extreme pressure from time to time, but it is particularly true of the ENG Editor working on a news programme. Most of the items will have been recorded that day, and the Editor has very little time to preview the tapes, identify the key elements of the stories in conjunction with the Reporters, and assemble news items of exactly the right length and content before the programme goes out.

A few ENG Editors, mainly at ITN, are expected to carry out maintenance on their equipment since for reasons of efficiency they sometimes work 'in the field' rather than at base and consequently do not have engineering support.

EFP production is increasingly used for location work in drama and light entertainment and it may use large tape reels or cassettes. Many 'Film' Editors now train in EFP.

The skills needed by an Editor working on film, video tape or PSC are somewhat difficult to define. A knowledge of television production techniques is essential as the Editor must know which shots will cut together and which will not. For example, two shots of the same moving car taken from either side will not cut together, as it will appear to the viewer as if the car is suddenly changing direction.

A degree of creativity is also essential since a cut which is made a few frames too early or too late can greatly diminish the aesthetic effect of a scene. In addition, creativity must be combined with ingenuity on the occasions when there seems to be no way that two shots can be cut together!

The ability to listen to other members of the team and understand and interpret their ideas is important, together with the ability to put one's own ideas into words.

An interest in current affairs and a broad general knowledge is a great advantage since in the course of their careers, Editors will work on a wide range of programmes. An understanding of the subject matter can be of great assistance in creating the right atmosphere in a programme.

Editors are usually assigned to a programme or a series, so that they have the satisfaction of following a job right through. They frequently work alone although several Editors may work on a very big production.

Editing is of course a long-term career in itself and with increased experience an Editor can hope to progress to ever more demanding programmes.

Applicants from outside the television industry who do not have previous experience can enter the editing field directly by applying for posts as Trainee Assistant Editors.

Typical Recruitment Profile

Trainee Assistant Editor

	Essential	Desirable
Physical		
Age	18+	19+
Eyesight	Good colour vision	
Hearing	Good	
Coordination	Excellent hand–eye coordination	
Education and Training	Broad general education to at least GCSE (grades A to C) standard	GCSE English (grades A to C) Adequate numeracy
Experience		Knowledge of film stocks and processing Knowledge of video tape production
Interests	Any aspect of photography or film making. TV and cinema. Current affairs and general knowledge	

	Essential	*Desirable*
Personality	Neat handwriting. Attention to detail. Well ordered mind. Observant. Conscientious. Able to work under pressure. Creative. Able to grasp technical matters. Able to listen to and understand the instructions and ideas of others	

ELECTRICIAN

Apprentices, House Electricians and Production Electricians

Electricians are employed by the ITV companies in a variety of jobs associated with the installation and maintenance of electrical facilities and the lighting of studios for programmes.

The task of the House Electrician is to install and maintain lighting and power supplies, etc., in offices and other company premises, and to supply power and light to studios and technical areas.

Production (or Lighting) Electricians, on the other hand, are employed to set up lamps in the studios. A typical television studio has lighting grids in the roof from which a variety of shapes and sizes of lamps are suspended. It is the job of Lighting Directors (see p. 111) to plan how the lamps should be arranged to create the desired effect for a particular programme. The Production Electricians use these plans to move the lamps along the grids and into the correct position. They are also responsible for ensuring that the lamps are fully operational. A good head for heights is needed for this job! Production Electricians are also employed in a similar function on some outside broadcasts and film locations although in these cases the lamps are usually free-standing or fixed to scaffolding or anything else that happens to be available.

In some companies the job of House Electrician and Production Electrician are totally separate but in others they are combined. Electricians may be recruited as House Electricians, Production Electricians or both, and where the jobs are separate, staff may sometimes move from one to the other after a period of time.

Vacancies usually occur for fully qualified Electricians, but occasionally apprenticeships are offered. On completion of the apprenticeship (which combines periods at college with on-the-job training), apprentices become fully qualified JIB approved Electricians. Promotion can be to Technician, Chargehand, Supervisor or alternatively to Console Operator. (See section on Lighting Director.)

All Production Electricians must be prepared to work irregular hours and House Electricians must work shifts. All are occasionally on call at irregular times. Production Electricians are also required to work away from home, sometimes under difficult conditions with a minimum of supporting facilities.

Typical Recruitment Profile

Apprentice Electrician

	Essential	*Desirable*
Physical		
Age		16–17 (unless redundant from another scheme)
Health	Normal fitness	
Eyesight	Normal colour vision	
Coordination	Good hand–eye coordination	
Education	Should have studied to GCSE standard in at least four subjects including Maths, English, and either Physics or Technical Drawing	

	Essential	*Desirable*
Interests		Electronics
Personality	Able to use initiative and to work as a member of a team. Willing to undertake further study	
Availability	Able to work shifts or irregular hours	

Note: Companies prefer to recruit apprentices from local schools and job centres rather than from outside the area. When vacancies occur, they will be advertised locally.

ELECTRONICS ENGINEERS AND TECHNICIANS

In many ITV companies, the traditional job descriptions for engineers are changing with developing technology and changes in operational practice. Some operational job functions are being combined, while in other areas such as VTR (video tape recording) the emphasis is changing from technical expertise to a greater degree of creative and artistic appreciation.

Television is a medium which relies for its existence on the skills of the Electronics Engineer. This section describes the work carried out in television broadcasting by engineers and technicians other than those engaged exclusively on sound.

In the present rapidly changing environment it is difficult to give precise job descriptions applicable to engineering posts within all the ITV companies. In many of the larger companies there is a division between the engineering and technical operations. In such cases, the engineering function is responsible for carrying out planning, installation and maintenance, while the technical operations function is responsible for the day-to-day operation of broadcast equipment. Other companies have no such division and as a result may require operational staff to carry out major maintenance work on equipment

Electronics Engineer conducting tests on a studio camera (STV)

and indeed engineering planning and installation projects.

Entry qualifications do, therefore, vary from company to company and potential employees are advised to check with individual companies before applying for posts.

The main areas of employment for engineers and technicians are shown below. These are not intended to be specialist functions and in most companies individual engineers will work in most, if not all, these areas during their careers.

Vision Control

In the suite of control rooms associated with each studio is an operational desk controlling the colour balance and exposure of each of the studio cameras. The engineer or technician allocated to this duty ensures that each 'shot' is correctly adjusted, and liaises closely with the Lighting Director (usually in the same control room) to ensure that the video output from the studio is both technically and artistically satisfactory. In some studio centres, video tape recorders are also controlled from this operational position.

The vision control staff may carry out maintenance on cameras, while in other companies this responsibility is delegated to separate central apparatus room staff or a maintenance department.

Video Tape Recording (VTR or VT)

The VTR section is a prime component of any television centre, and carries a high operational workload. The section has two main functions.

The first of these is to edit programmes which have been recorded on video tape. A separate section of the book describes the role of the Editor in detail.

The second is to record programmes as they are made in the studio, replay recorded inserts into live programmes, replay programmes for transmission, and to 'time shift' programmes which may be required to be transmitted at different times throughout the United Kingdom.

Staff who join the VT department are normally concerned with recording and replaying programmes. Those with creative ability may progress into editing as opportunities arise.

In some companies, VT staff are also responsible for the maintenance of their equipment.

Telecine

Although most television programmes are made and transmitted on video tape, a sizeable minority are made on film. In addition, many cinema movies are of course shown on television. The telecine machine is designed to convert film into a television signal for transmission which it does by 'scanning' each shot. The task of the operational engineer or technician in the telecine department is to obtain the most satisfactory audio and video output from the equipment.

Cine film is a medium far from ideally suited to reproduction on television due to its wide contrast range and varying colour rendition. Telecine staff can correct these aberrations with sophisticated electronic processors. The need to cope with a variety of film gauges,

Telecine Technician (Granada)

cinemascope, widescreen and stereo sound tracks adds complexity and interest to this operation.

As in VTR, the responsibility for the maintenance and technical alignment of the equipment may lie within the department, or with a separate maintenance section.

Central Apparatus Room (CAR)

The central apparatus room is the nerve centre of the studio complex. In it is housed a large proportion of the electronics providing routing of video and audio signals throughout the station. Much of the video mixing, processing and digital effects equipment is also housed here, as is the equipment providing central timing and test signals.

This area is the communications hub, between areas within the studio complex, to other ITV companies, to the transmitters and to British Telecom who provide communication lines.

Engineers working in this area are responsible for the monitoring and, where necessary, the correction of the technical parameters of all audio and video sources within the complex. In particular, they are responsible for incoming signals from other centres and the routing of those signals to the transmitter.

Outside Broadcasts

Many television programmes are transmitted from, or recorded at, locations away from the studio centres. The control rooms and electronic equipment used in these circumstances are fitted within 'outside broadcast vehicles' sometimes known as 'scanners'.

The engineers employed with these scanners perform similar duties to those already mentioned. However, the absence of base support and the need to work under frequently difficult environmental conditions often place greater emphasis on the individual's initiative and motivation.

The job is more varied than those in a studio environment and many staff derive greater job satisfaction from this type of work.

Radio Links

When it is necessary to transmit an outside broadcast live from location, circuits for vision, music and communications are required back to the studio centre. Occasionally, permanent landlines are available, but in most instances the vision and music circuits are carried by a temporary microwave radio link.

A typical link would consist of a vehicle adjacent to the outside broadcasting location, housing a microwave transmitter and communications equipment. The aerial for this transmitter would be mounted either at the top of the vehicle's mast, on the vehicle roof, or dismounted from the vehicle and rigged on a suitable roof top.

The 'receive' point for the link may be a permanently installed aerial located on a transmitter mast or high building. Alternatively it may be another links vehicle on a remote hilltop which receives and re-transmits (mid-point) the signal to a further receive point, or feeds the signals into the permanent ITV lines network.

The microwave frequencies used for these links require unobstructed 'line of sight' transmission paths to ensure that satisfactory circuits are provided for the broadcast. Ideally the route should be carefully surveyed by the radio links engineers in advance, to check for obstructions such as hills or high buildings, or other problems. There are times, particularly in news operations, when links are required at very short notice from locations which have not previously been used or surveyed. On these occasions the engineer must use initiative and experience to establish a usable link.

Engineers working on radio links are required to rig equipment under adverse weather conditions and often at heights. Considerable self-reliance, resourcefulness and determination are called for.

Maintenance Department

All studio centres have an electronic maintenance workshop where repairs and routine servicing of equipment are carried out.

Advances in microtechnology have resulted in more complex equipment with built-in automatic correction circuits, minimising the need to recruit operational staff with a technical background.

As a result, operational engineers in some companies are now only responsible for first-line maintenance, with the maintenance department of specialist engineers undertaking the major technical workload.

In addition to the routine maintenance and repair of broadcast equipment, such as electronic newsgathering cameras, computer graphics, video tape recorders, edit controllers, etc., some maintenance sections undertake a varied amount of development and design work as well as evaluation and acceptance testing of equipment prior to purchase.

Planning/Installation/Research and Development

This section is responsible for the detailed planning of new technical projects to provide the facilities required by the operational staff. The work requires expertise outside traditional electronics skills, since there is a need to understand the difficulties and limitations imposed by building and air conditioning requirements. An understanding of project management and budgeting is also essential.

Once the planning is completed the section is responsible for carrying out the installation. If outside contractors are used, the section will supervise their work. Considerable time is spent on acceptance tests of new equipment and systems, a duty which is often shared with engineers from the maintenance sections.

In addition to planning and installation, this section is often heavily involved in the assessment of new equipment or systems prior to a purchase decision being made.

In some of the smaller companies this section is incorporated within the maintenance department.

Career Structure for Engineers and Technicians

Entry is normally at trainee or first-year engineer grade. There is automatic progression over six years to the grade of substantive engineer. This period and grade can vary from company to company and able candidates may progress faster.

Thereafter, promotion is by merit and competition through two or more grades to a senior supervisory grade engineer in charge of the technical staff and facilities at, say, a studio, an OB unit, or a group of technical facilities departments. In most companies, engineers at substantive grade and above will be skilled in many or all of the jobs described. Some companies, however, prefer their staff to specialise in a small number of skills.

The following recruitment profiles reflect the jobs to which applicants with differing academic qualifications and aptitudes may aspire.

Typical Recruitment Profile

Trainee Engineer: Maintenance

	Essential	Desirable
Physical		
Age	18 +	22 +
Eyesight	Normal colour vision	
Coordination	Good hand–eye coordination	
Education	(1) BTEC Higher Cert/Diploma Electronics with TV or communications, or (2) City & Guilds T5, or (3) HNC/HND in Engineering or Electronics or Telecommunications	Degree in Engineering or Electronics or Telecommunications
Interests	Electronic and non-specific wide range of technical teamwork, and media-related interests	Ability to demonstrate practical aptitude
Personality	Self-reliant, but able to work happily as a member of a team	
Availability	Willingness to work irregular hours and/or shift work over seven days	

Typical Recruitment Profile

Trainee Engineer: Planning, Research, Installation and Development

	Essential	*Desirable*
Physical		
Age	21 +	
Eyesight	Normal colour vision	
Coordination	Good hand–eye coordination	
Education	Degree in Engineering or Electronics or Telecommuni-cations. (Applicants wishing to follow a career in projects will be expected to gain initial experience in other engineering departments.)	(1) Hons Degree in Engineering, or Electronics or Telecommuni-cations. (2) Project Management awareness. (3) Financial Management awareness. (4) Technical report writing ability
Interests	Electronic projects and a wide range of technical, teamwork and media-related interests	
Personality	Equable temperament, able to work happily as a member of a team	
Availability	Willing to work irregular hours and/or shiftwork over seven days	

Typical Recruitment Profile

Trainee Technical Operator: Other Areas

	Essential	Desirable
Physical		
Age	18 +	21 +
Health	Good stamina	
Eyesight	Normal colour vision	
Coordination	Quick, calm ear–hand reactions	
Speech	Clear	
Education and Training	Broad general education to at least GCSE standard (grades A to C) including Maths and Physics to 'A' level	BTEC/City & Guilds/ONC/OND in Electronics, Computer science or Communications
Interests	Wide range of technical team work and media-related interests	Able to demonstrate artistic aptitude
Personality	Self-reliant but able to work happily as a member of a team	
Availability	Willingess to work irregular hours and/or shift work over seven days	

FLOOR MANAGER

■■ *Including Floor Assistant and Stage Manager*

The responsibilities of the Floor Manager fall into two main areas: (1) liaison between the Programme Director and the studio floor, and (2) management of the studio floor.

It is of course the Programme Director who has overall responsibility for how a programme is made. During studio rehearsals and the final recording, however, the Director is seated in a control room which is usually separated from the studio by a glass screen, and views the proceedings on a bank of television monitors. The Floor Manager receives instructions from the Director via headphones, and takes action accordingly. The Director may for example decide during studio rehearsals that an actor should move to another position, or speak his lines in a different manner. It is the Floor Manager's responsibility to pass on these instructions in a way that will achieve results. This often demands a great understanding of human nature, since the actor may already be under considerable pressure, or may have different ideas from the Director about how the part should be played. Similarly, Floor Managers may have to coax nervous members of the public into talking on local magazine programmes. They also give performance cues to the performers and provide prompting where necessary.

Floor Managers will work with many different Directors over a period of time and should be able to understand the way each one likes to work and anticipate their wishes.

Floor Managers are also the coordinators and managers of all that is happening on the studio floor. They not only make sure that artists are ready when needed, and that any extras and walk-on artists know where to stand and what to do, but also ensure that all the other aspects of the production are ready when required. Props must be in position, cameras and microphone booms must be correctly positioned according to the plans, make-up must be completed on time and a host of other items must be checked. Once again, this aspect of the job demands considerable interpersonal skill since the members of the production team are all experts in their own fields. It is important that Floor Managers are able to command the respect of the

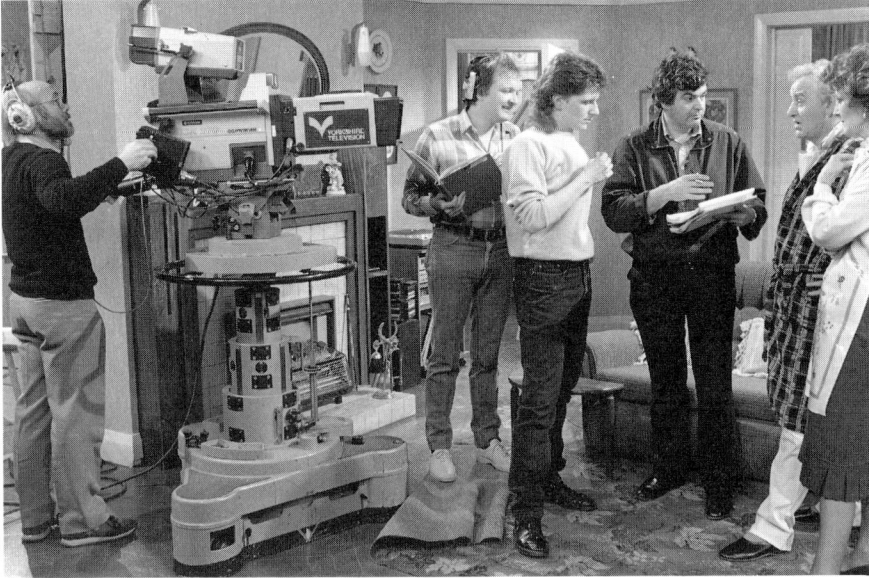

Floor Manager (second from left) and Director in discussion with artists during rehearsal (YTV)

production team even though they may not have the technical knowledge that the team members have.

If there is an invited audience, the Floor Manager is responsible for their safety, well-being, and involvement in the programme.

The amount of responsibility that Floor Managers are given depends upon the Director, and varies between companies. Some may be given the task of deciding the order in which the scenes are to be shot. Some may also be responsible for drawing up the final script (i.e. the total plan of action for sound, cameras, artists, etc.). In many companies, however, these tasks are not carried out by the Floor Manager.

The Floor Manager is present at outside rehearsals (which are held mainly for drama and some light entertainment productions). These are the early rehearsals which are usually held in large public halls away from the studio. The Floor Manager is mainly a bystander at this stage, taking the opportunity to learn the production very thoroughly, and making constructive suggestions to the Director about how things could be improved.

Floor Managers also work on outside broadcasts, carrying out much the same function as in the studio.

In many companies, Floor Managers are also involved with film production in the capacity of Assistant Directors. The responsibilities and techniques are different from those in video tape production. First Assistant Directors have wide authority in many areas.

In the pre-production period the First Assistant will work with the Production Manager to 'break down' the shooting script and examine the likely requirements and responsibilities. The 'First' will also work closely with the Director to find out what he/she has in mind and how the whole film is visualised, and will then develop a very detailed shooting schedule. Additionally, it is the 'First's' responsibility to coordinate any special effects, stunts, and casting, etc. An awareness of health and safety in the working environment is essential, including knowledge of the relevant legislation and employer requirements.

The job is very demanding both physically and mentally. The Floor Manager must have a total understanding of the production, and complete commitment to it. At the same time, the job calls for long, irregular working hours, and very little opportunity to sit down for a rest. An equable temperament is necessary to cope with the crises that may occur, particularly on programmes which are live rather than recorded. It is easy to see that academic qualifications are less important for the job than the right personality and plenty of enthusiasm.

In the past many Floor Managers have started their careers as Floor Assistants, which is the equivalent of a runner (or 3rd or 4th assistant) in the film industry. The minimum recruitment age for Floor Assistants is usually 18. Floor Assistants are an essential part of the production team, particularly if a large drama is in production. However, very few are now employed in ITV. Their role is to make sure that performers are in the right place at the right time, and that they are made comfortable at all times.

Today Trainee Floor Managers may be recruited from a wide variety of jobs in the television industry. They are rarely recruited from outside unless they have similar experience in the theatre. Progression is from trainee to 1st-, 2nd- and 3rd-year Floor Manager and then substantive grade. Junior staff usually assist more experienced Floor Managers, and may work on their own on short, straightforward programmes. The top rung of the promotion ladder is Senior Floor Manager.

Stage Manager

The role of the Stage Manager is to organise outside rehearsals including the arranging of rehearsal rooms, ordering and moving of rehearsal props, keeping the script up-to-date during rehearsals, prompting the performers, ensuring continuity of props, marking out the floor of the rehearsal room, etc.

All Stage Managers have extensive experience in television or the theatre. A few are employed on a permanent basis, but most are freelance and are employed on short-term contract for a particular production. (A 2nd Assistant in the film industry does roughly the same as a Stage Manager on a video production.)

Typical Recruitment Profile

Trainee Floor Manager

	Essential	*Desirable*
Physical		
Age	24 +	
Health	Good stamina, agility	
Hearing	Good (able to hear clearly over headphones)	
Speech	Clear	
Education and Training	Broad general education to at least GCSE grades A to C standard	GCSE grades A to C in a range of subjects
Experience	Experience of handling large groups of people in stressful situations	Experience of working with actors in theatre, film or TV General experience of the business including technical aspects of television

	Essential	Desirable
Interests	Drama	Musical appreciation and an understanding of musical notation Any aspect of film, television and theatre
Personality	Good working command of English. Confident and genial personality. Able to understand and handle all types of people. Sense of humour. Self-discipline and initiative. Ability to grasp technical matters. Unflappable. Tactful. Sensitive to performance. Dedication to final product. Leadership. Ability to work as a member of a team	
Availability	Able to work long, irregular hours at any location	

GRAPHIC DESIGNER

Graphic Designers are normally employed within the Design Department and usually work alongside the Set Designers. Their work is most frequently seen in the opening and closing titles of a program-

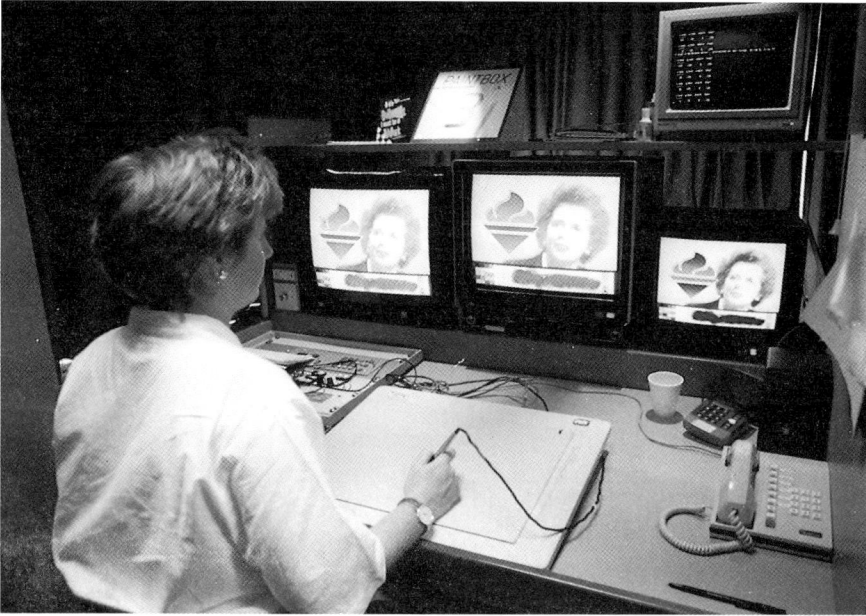

Graphic Designer using the 'Paintbox' (ITN)

me, but they can be called upon to undertake a host of other tasks including the preparation of cartoon sequences, weather charts, economic forecast charts and even such things as old bank notes, letters, driving licences, paintings, etc., for use as props in drama programmes. The range of tasks is endless and the job calls for considerable ingenuity and imagination.

The simplest opening titles consist of the name of the programme superimposed on the first scene, in which case the Graphic Designer is responsible for the layout and the lettering. More complex opening titles might include moving cartoon sequences or photographs and once again the Graphic Designer is responsible for preparing the constituent parts. The wishes of the Programme Director must of course be taken into account, as must the wishes of the Set Designer and other members of the production team. There must be continuity of theme and mood between graphics and sets, and both must reflect the atmosphere of the programme if they are to be effective.

Some of the work of the Graphic Designer is done by hand or with printing equipment in the traditional way, but a great deal is done on computers. Computer graphic systems such as 'Paintbox' enable the Graphic Designer to experiment in ways which are not possible using the more traditional methods. Suppose, for example, that the Graphic

Designer 'paints' a picture of a girl in an apple orchard, but is then unhappy with the colour and layout. The colour of the apples, or anything else, can be changed instantly and the girl can be moved from her original position to another point in the picture to produce the desired effect.

It is usually impossible to say which of the graphics you see on your television screen are produced by hand and which are produced by computer. The three-dimensional moving sequences, however, have almost certainly been produced on a computer.

It is important to remember that whichever way graphics are produced, the creativity must come from the Graphic Designer.

One area of graphics which is not normally the responsibility of the Graphic Designer is the production of captions showing, for example, the name of someone being interviewed for a news item or a sports programme. These captions are produced by the Caption Generator Operator using a machine similar to a word processor.

The job of the Graphic Designer calls for a wide range of skills, and excellent drawing ability combined with a high degree of creativity is obviously essential. The ability to communicate ideas and to be sensitive to the ideas of others is also essential. A wide range of interests and an enquiring mind are very useful as the Graphic Designer can be called upon to work on any kind of programme.

Recruitment is normally as a Trainee Graphic Designer or Technician, and newcomers will usually begin by assisting more experienced staff on simple graphics. Promotion through several grades to Senior Graphic Designer involves increasing responsibility for more prestigious and complex productions.

Typical Recruitment Profile

Trainee Graphic Designer

	Essential	Desirable
Physical		
Age	21 +	
Eyesight	Good colour vision. Able to do fine, detailed work	
Coordination	Excellent hand–eye coordination	
Education and Training	Degree in Graphic Design or Licenciate of Society of Industrial Designers	Some training in Fine Art
Experience		Commercial art studio. Knowledge of photographic techniques Computer operations
Interests	Painting/drawing/graphics Contemporary design	Photography or other creative hobbies. Theatre, film, television
Personality	Ability to communicate ideas. Ability to work as a member of a team. Highly creative. Able to work under pressure at times	
Availability	Willing to work long hours if required	

JOURNALIST

▓ *Including Reporters/Correspondents/News Readers/Editors/Writers*

The term 'journalist' is applied to a number of different grades of staff all requiring a journalistic background.

Journalists are employed both by Independent Television News (ITN) and the ITV programme companies on news programmes. Other Journalists may work on news or current affairs programmes, and some are employed to prepare stories for ITV's teletext system, Oracle.

ITN has the responsibility for providing the national news coverage to the network on 'News at One', 'News at 5.45', and 'News at Ten'. It also provides the news on Channel 4 and many other bulletins and news programmes. The extent of its operations means that ITN employs almost as many Journalists as the rest of the network put together.

Journalists in the ITV companies provide regional news and magazine programmes for their own regions. Sometimes they feed regional stories into ITN, but generally ITN sends its own staff to cover assignments throughout the country. Journalists in the programme companies have considerable variation in their jobs, sometimes reporting, otherwise working as Production Journalists in the newsroom. At ITN, the work is structured far more rigidly.

Any Journalist who works in television news must expect to operate under considerable pressure. News programmes, unlike most others, are live and there are therefore strict deadlines to be met. There is also very little preparation time, since news is only of interest if it is fresh.

Newly appointed Journalists at ITN can expect to spend most of their time behind the scenes, searching for stories and writing news scripts. In the programme companies, newcomers may be sent on reporting assignments early on, particularly if the Journalist has broadcasting experience, but there will still be plenty of backroom work to do. Promotion to senior posts in both cases is on merit.

▓ *Sources of News*

Potential stories come from a wide variety of sources. All of the programme companies have extensive local contacts in their own areas, such as local politicians, police, etc., and it is of course important that Journalists are able to develop a good relationship with such people. They also rely on 'stringers', who are freelance journalists based in a particular area who feed stories to several news organisations at once. Some stories may be followed up from newspaper or magazine cuttings, and many are 'diary' stories, i.e. they are known in advance, such as sports meetings and royal visits. Many ITV companies operate small, regional news centres in major towns and cities.

There are several news agencies to which companies may subscribe. Some programme companies take Press Association services for domestic news. ITN subscribes to several news agencies and relies quite heavily on the wire services they provide, especially for overseas news. ITN has its own contacts and stringers in the same way as the programme companies, and also has its own specialist correspondents to provide news stories.

Reporters

Once the Editor (or in the case of ITN, the Duty News Editor) has decided which news items should be followed up, a Reporter may be despatched to cover the story. If, for example, there is a strike at a local factory, the Reporter will go to the scene with a technician or technicians to operate the camera, record the sound, and light the scene if necessary. The Reporter's task is to assess the situation on the spot, decide on the best way to present the story, interview key people such as the Manager and Shop Steward and perhaps prepare a piece in which the Reporter talks directly to the camera, telling the viewer the background to the story. This is usually done out of sequence, and so notes must be kept to enable the Video Tape Editor (not to be confused with the Editor of the programme) to assemble the item in the correct order for viewing. Reporters must be able to work on their own initiative as there is no Programme Director present to tell them what shots to take, or Scriptwriter to tell them what to say. They must have the social skills to win the confidence of people ranging from old-age pensioners to film stars and politicians. They must be able to formulate the right questions and commentaries on the spot, in a language that will appeal to both educated and uneducated members of the community.

They must also be prepared for frustrations, as what appears to be a good story can be cut from the programme at the last minute if a bigger story comes along. Since news is a highly perishable commodity the Reporter's work is often in vain as an item can rarely be used in a later programme.

Correspondents

At ITN and some of the larger programme companies, there are a few specialist Reporters who cover subjects which need an in-depth knowledge. There are for example Political, Diplomatic and Industrial Correspondents. Foreign Correspondents for ITN may be based in cities such as Washington and Moscow where there is sufficient news to justify a permanent presence. They become familiar with the local way of life and are accepted by the local community. They send back regular reports, and also report on any major news items that occur in their area. Most of the programme companies do not employ Correspondents.

Foreign Correspondent Simon Cole reporting on the Kuwaiti Hijack, 1988 (ITN)

News Readers/News Presenters/Newscasters

News Readers and Presenters tend to be the personalities of television news. They are usually experienced Journalists. They present news items from the studio and act as anchor men and women, linking and introducing items from various Reporters. Their own scripts may be prepared for them by other Journalists, or they may write their own. They must do a great deal of preparation in a very short time before the programme, making sure they are thoroughly familiar with all the planned news items.

They must also be aware of any developments which mean that last minute changes may be made. News programmes are of course live, and News Readers must therefore be able to cope with any eventuality in a calm, professional manner. They are selected on the basis of their experience in television, appearance, speech and personality.

Working Behind the Scenes

There are a number of other jobs in television news which are carried out by Journalists but which are not visible to the general public.

Scriptwriters, for example, are concerned with writing news stories. Their other responsibilities usually include liaising with Reporters, and technical and production staff on the editing of video tape; organisation of maps, electronic graphics and captions, and scripting the final item to be transmitted.

Editors may be involved in a variety of tasks. In the programme companies, for example, the Programme Editor or Executive Editor has the key managerial role for the local news programme. The job usually includes such things as making the final decision on content, vetting items for accuracy, libel, etc., as well as all aspects of planning, budget control, preparation, scheduling, management of staff, and ensuring that all deadlines are met. In a small regional company, one person may be responsible for all of these tasks whereas in larger companies there is a hierarchy of staff, including News Editor, Production Editor and Bulletin Editor among whom the tasks are divided. In broad terms, the title 'Editor' is given to staff whose job contributes to the overall assembly and preparation of the news programme.

At ITN there are a number of staff involved in the editorial process including Programme Editor, Chief Sub-Editor, Senior News Editors, Duty News Editors, Deputy News Editors, etc. All of these are quite senior positions and are only reached after several years' experience.

Other Posts for Journalists

Journalists are employed as Promotion Scriptwriters, and their task is to produce programme trailers and to prepare scripts for the Continuity Announcers who appear between the programmes.

Press Officers are employed as part of the Public Relations function of the companies. Their role is to report on occurrences within the television companies which are of interest to people outside the organisation. They may, for example, prepare press releases to promote a new programme, or to announce an organisational change.

Typical Recruitment Profile

Trainee Journalist (Programme Companies)

	Essential	Desirable
Physical		
Age	21+	
Health	Good stamina	
Speech	Good, clear speech	
Appearance	Likely to be acceptable to all types of people	
Education and Training	Two good 'A' level passes, preferably including English	Degree, preferably in English or a subject related to Communications, but the choice of subject is not critical
Experience	About 2 years' experience in press or radio journalism	

	Essential	*Desirable*
Interests	Current affairs; a wide-ranging taste in reading; writing; all aspects of the media; social interests	
Personality	Self-confidence, leadership, drive, maturity, attractive personality, able to work as a member of a team, and to make quick decisions. Highly articulate, able to remain calm under pressure, enquiring mind, self-reliant, a good command of written and spoken English. Able to win the confidence of all types of people	
Availability	Willing to work irregular hours at any location	

Note: From time to time some of the programme companies run formal training schemes for Trainee Journalists without previous experience in the press or radio. These schemes normally begin in September but they are not run every year as much will depend on the company's staffing needs. They are usually advertised in the national press. Competition for vacancies is very tough.

▮ *Journalists (ITN)*

Most journalists at ITN are recruited from the programme companies where they will be expected to have gained a good grounding in television news.

In most years, however, there is a very small annual intake of graduates for the Graduate Editorial Trainee Scheme. Applicants are expected to have a good class of degree, and to be able to demonstrate a natural flair for writing news. Previous professional experience in journalism is not essential. The personal qualities mentioned above are equally applicable to candidates for this scheme. Competition for the posts is very tough.

LIBRARIAN

The ITV companies have differing approaches to library services and a variety of job opportunities exists within the industry. Some companies employ Librarians in a general capacity to deal with a number of information sources. Others divide their library service into separate sections such as reference, film and video tape, news information, stills and music with perhaps a library manager in overall charge. While each of these roles calls for certain specialist skills, they all demand at least some understanding of the techniques of television production and an awareness of the deadlines to which programme makers are forced to work.

Most Librarians within ITV are employed in film and video tape libraries handling a variety of broadcast materials for which a level of technical understanding is required. Professional library qualifications are becoming essential together with some study of non-book materials as part of the course. As the use of computer systems expands, Librarians in the television industry increasingly need to be computer literate. They also need to decide whether to develop library systems within their own companies or to access external commercial databases.

As libraries develop within television, the Librarian's role is expanding. They are no longer merely the custodians of various cans of film and cases of video and audio tape but must be able to exploit the information held on the material. ITV library staff are increasingly able to contribute creatively to programme-making by informing researchers, journalists and directors of what material is available to them, and advising them how to exploit the collection successfully. If a request cannot be fulfilled within the company, Librarians should be able to point the enquirer in the right direction, either within the

ITV network or the industry generally. After a relatively late start in some ITV companies, libraries are now rapidly developing new systems of cataloguing and classification to cope with information handling in an expanding and changing industry.

The television industry is becoming aware that much of its output has a value beyond fulfilling day to day transmission requirements, and as a result, some companies employ archivists either as part of the library structure or in a separate post to review library holdings and ensure that material of archive importance is permanently retained. Archivists are often involved in liaising with researchers from outside the television environment, for example academic institutions.

Library staff should be capable of working under pressure to meet transmission deadlines. Accuracy of work is also necessary as Librarians play an integral role in making sure the correct material is available for editing and transmission.

Those seeking employment in this area should not have any illusions that the job is a quiet backwater. Life can become pretty hectic when transmission deadlines approach, particularly in the news environment. The advantage of this, however, is that there is rarely a dull moment and the job can be very rewarding compared with more conventional library work.

Typical Recruitment Profile

Trainee Assistant Librarian

	Essential	*Desirable*
Physical		
Age	18 +	21 +
Health	Good physical stamina	
Eyesight	Normal colour vision	
Speech	Clear, good telephone manner	
Education and Training	Broad general education usually to degree standard	Degree or postgraduate diploma in librarianship preferably with

	Essential	*Desirable*
Experience		non-book materials option Some experience of a broadcast industry library perhaps on placement as part of a degree course
Interests	Broad, including current affairs, television production and film making. An interest in all kinds of music is essential for Music Librarians	
Personality	Outgoing, able to operate calmly under pressure to meet production deadlines. Capable of dealing tactfully and efficiently with a variety of people both face to face and over the telephone. Able to make a creative contribution to programme making	
Availability	Hours can be irregular	

LIGHTING DIRECTOR

In order to understand the role of the Lighting Director, it is important to appreciate the difference between illumination and lighting. Illumination gives the basic light necessary for the camera to record the picture. Lighting adds to the quality of the picture and can play a large part in creating the right sense of mood and atmosphere. It is also used to add to the illusion that, for example, a television studio is really a living room in a suburban house. Carefully planned lights can make the viewer feel that daylight is streaming in through a window or a front door. The possibilities for the creative use of light are endless and the job can be highly satisfying, even if it is largely taken for granted by the viewer.

Newly appointed Lighting Directors (usually called Lighting Assistants) are allocated to programmes which require fairly straight-forward lighting and their work is supervised by experienced Lighting Directors. A news programme, for example, is relatively simple to light as the news presenter remains seated, and is usually seen against a single, fairly simple background. The task for the Lighting Director is to give the face shape in the most flattering way, and to separate the artist (News Reader) from the background, thereby creating a three-dimensional feeling to the picture. The technique is very similar to that used by a portrait photographer.

At the other end of the scale, a Shakespearean drama, for example, will present many challenges. Not only are the many characters moving around, making it difficult to light their faces perfectly, but lighting has also to be used to create the feeling of dark castles, light streaming from behind pillars, long corridors, and candle light. Clearly, the Lighting Director must work very closely with the Set Designer since poor lighting can ruin the effect of a well designed set.

All members of the production team will meet early on to plan the production and discuss any conflicting interests. This discussion will enable the Lighting Director to decide how to light the production in a way which will avoid, for example, the lights casting a shadow of the sound boom across the picture.

Electricians then set to work, positioning the lights in the studio in accordance with a chart prepared by the Lighting Director, and under his/her supervision. During the recording of the programme, the brilliance of the lights can be altered as planned from a lighting console in the studio control room. This is done by the Lighting

Console Operator who is an Electrician.

Most of the lights in the studio are fixed to a grid high in the ceiling, and can be moved around with ease. If the Lighting Director is working on an outside broadcast, however, there are many additional problems to be faced. Natural daylight is not particularly suitable for television, and although in programmes such as football matches there is little alternative, a better quality of light is needed for drama and light entertainment. The Lighting Director must therefore have a supply of cumbersome lamps to transport to the location. If a scene is to be shot inside a house, for example, it can be very difficult to place the lamps in the ideal position and the Lighting Director has to resort to ingenuity even more than usual.

Some television programmes which are made away from the studio are on film rather than video and in these cases, it is the Lighting Cameraman rather than the Lighting Director who is responsible for lighting. The Lighting Cameraman is the most senior member of the film camera crew.

Lighting Directors are recruited from a variety of sources and it is difficult to define a typical career path. They are usually recruited from among the existing staff within the company and are only recruited from outside if they already have similar professional experience in film or theatre lighting.

The job requires a mixture of technical knowledge and creativity and for this reason, many Lighting Directors have previously worked on camera crews. Some have been engineers, and a few have been electricians but almost any related jobs in television could be a starting point. The technical aspects of the job are mainly concerned with the physics of light and electricity. A high degree of perception, imagination and creativity is also needed.

Typical Recruitment Profile

Trainee Lighting Director (Lighting Assistant)

	Essential	Desirable
Physical		
Age	18+	21+
Health	Good stamina and agility	
Eyesight	Good colour vision	
Education and Training	Broad general education to at least GCSE grade A to C standard	GCSE grades A to C Physics, Maths, Art. 'A' levels (subjects not specified)
Experience	Several years' experience in a related department in television	
Interests	Creative, artistic interests	Photography or film making, art and design. Theatre, television and films
Personality	Ability to communicate ideas and work as a member of a team. Highly creative. Able to work under pressure at times. Able to supervise staff from other disciplines	
Availability	Willing to travel and work unsocial hours when required	

LOCATION MANAGER

It is the task of the Location Manager to seek out and investigate suitable locations for use in the shooting of programmes and, most notably, dramas.

The starting point for the search is of course the script, and the Location Manager will discuss with the Producer and Director the kind of location which will most closely meet their interpretation of the script.

Although it is important to find, for example, a house of the correct historical period or a landscape with the right atmosphere for the drama, there are many other factors which must also be taken into consideration. Ease of access and parking for support vehicles, catering vehicles, crew cars, artists' cars and vehicles needed in the action must be considered. Any inconvenience that is likely to be caused to the owners or users of the property must be anticipated, and steps taken to minimise any disruption.

External factors must also be taken into account, for example noise from aircraft at nearby airfields or traffic on motorways. Even the tidal rise and fall on a river estuary can have an important influence on its use as a location as the continuity of shots taken out of sequence, or over a period of time, will have to be borne in mind.

It is the Location Manager's responsibility to control any factors which can be controlled, and where necessary to negotiate with the Police, local authorities, military authorities, etc., to stop traffic and limit flying, and so on, while shooting takes place. Any factors which cannot be controlled must be taken into account when planning the shooting schedule.

The Location Manager may take a series of photographs to illustrate the proposed locations, and they will be discussed by the production team. If the locations appear suitable, the production team will visit them to ensure that all their varying needs are met.

Once the locations have been chosen, the Location Manager is usually responsible for all the preparatory administration and paper-work. This may include, for example, discussing terms and conditions for contracts with land-owners for the use of their property. The contracts will then be drawn up by the Contracts Department. There will be many letters to write, confirming various arrangements with local authorities, etc., and with location caterers who supply meals to the crew and artists during the shoot. There will also be paperwork to be completed during the shoot and during 'clear-up'. In some companies, much of the paperwork may be done by the Production Assistant rather than the Location Manager.

Location Managers must have a sound business sense and the ability to negotiate effectively with officials and members of the public. An ability to get on well with all types of people is essential, together with a thorough knowledge of the requirements of programme production. Location Managers must be perceptive, have enquiring minds, and be able to interpret the Director's ideas when finding suitable locations. Vacancies are normally filled from within the organisation, or by applicants with similar experience in the film or video industries.

No recruitment profile is given for this job.

MAKE-UP ARTIST

The television viewer is likely to be aware of the work of the Make-up Artist only when watching a drama in which, for example, a character ages over a number of episodes, or a light entertainment show in which exotic fantasy make-up is used for the dancers. In such cases as these, make-up is used to transform the performer, either in order to develop a character or to add glamour.

Much of the work of the Make-up Artist is, however, 'corrective'. This involves the neatening of hair, the application of powder to shiny noses and foreheads to prevent reflections from the studio lights, and the general tidying of the appearance of a person who is about to appear in front of the camera. Corrective make-up forms by far the greatest part of the job for Make-up Artists who are employed in the smaller television companies. The programmes produced by such companies are mainly news and current affairs, or perhaps quiz programmes in which transformation of appearance is not necessary.

Programmes such as period dramas, science fiction series and light entertainment can present many challenges to Make-up Artists. They may be called upon not only to apply facial make-up but also body make-up for dancers, and to style hairpieces and moustaches, etc.

Experienced Make-up Artists may occasionally work with latex foam or other materials to change the shape of a face, or they may be required to create all kinds of scars and realistic wounds. Such special effects take time, ingenuity, imagination and skill to achieve. Prog-

Small World 1987, Make-up Artist (Granada)

ramme Directors decide what general effect is called for, but Make-up Artists must be able to interpret their ideas. They must also liaise with Set Designers, the Costume Department and the Lighting Director to ensure that continuity of style is maintained.

There is, of course, a much less glamorous side to the job. All members of the Department are responsible for the care and cleanliness of their own equipment, and the cleaning of wigs and other hairpieces is a tedious task. General attention to skin care is also important.

Hairdressing is a major part of the job, and Make-up Artists are frequently called upon to neaten hair, apply heated rollers and wash hair. They should also be able to cut and tint hair in period and modern styles. In many companies, the styling of wigs is an important aspect of the job. Every Make-up Department will therefore prefer to recruit trainees who have been previously trained in hairdressing.

The job can be very demanding both physically and mentally. The

Make-up Artist spends most of the day standing and bending over to apply make-up. This can cause considerable strain to the back and feet. The work is often carried out to a tight schedule, and during recording or filming the Make-up Artist must be on hand to touch up or alter make-up as necessary. If the programme is being made on location rather than in the studio, the job can involve standing around in the cold and wet and working without much in the way of back-up facilities.

It is essential that the Make-up Artist has the personality and the maturity to deal with all types of people. Many actors, actresses, politicians and members of the public are nervous before appearing in front of the camera. The Make-up Artist is often the last person they meet before making their appearance, and becomes the focus of their tension.

Entry into the job of Make-up Artist is usually as a trainee, with promotion to Make-up Assistant after two years and Make-up Artist after a further two years. Promotion beyond that level to Senior Make-up Artist and Supervisor is on merit. Junior members of staff will normally work as a member of a team on a particular production, under the leadership of a Senior Make-up Artist or a Supervisor. They will gradually take on more responsibility and perhaps work alone on fairly straightforward programmes.

After four or five years' experience, a Make-up Artist may take on full responsibility for a major production as well as supervision of teams of staff. Senior staff also become involved in research for programmes, such as discovering what type of make-up is appropriate to reflect a given period, or how a bullet wound might look.

Typical Recruitment Profile

Trainee Make-up Artist

	Essential	Desirable
Physical		
Age	21 +	
Health	Good stamina. Able to stand for long periods. Personal cleanliness	

	Essential	Desirable
Eyesight	Normal colour vision	
Coordination	Hand to eye	
Education and Training	City & Guilds, BTEC or other recognised qualification in Make-up and/or Hairdressing and/or Beauty Therapy *or* art school training in portraiture, sculpture or fine art	GCSE grades A to C in English, History and Art
Experience		Working with people. Experience of theatrical make-up on amateur or professional basis. Experience as a hairdresser or beautician
Interests	Creative activities, people	Television, drama, film, history
Personality	Mature, equable temperament, tactful, discreet, creative, artistic, good fashion sense, able to work as a member of a team	
Availability	Able to work irregular hours at any location	

MANAGEMENT

The role of Managers in television is, in most cases, very similar to their role in any industry. The practice of management begins with supervisors who, although not holding the title of Manager, are responsible for the first-line management of staff and other resources. At the middle management level are Heads of Department and at senior management level are Controllers, General Managers and Directors. At the top is the Managing Director. This is a rough guide and the number of levels of management varies according to the size of the company.

Some Managers are responsible for the work of large numbers of people while others, particularly those in specialist advisory areas, may have few or no staff reporting to them. They are, however, responsible for managing other resources such as time, money and equipment, and for achieving results through others.

There are a number of ways of becoming a Manager in an ITV company. Many of the departments are unique to the industry, particularly those involved in the making of programmes. Most Managers in these departments have worked their way up from the bottom, since it is a great advantage for them to have a thorough knowledge of how television programmes are made. Others may have gained similar experience in another television company and moved across as opportunities occurred.

Examples of management posts which are unique to the industry are the Executive Producer of a programme and the Managing Editor of a news bulletin. Heads of Department and Controllers of production, operational and most technical areas will typically have begun their careers in television.

Some Managers are, however, recruited from outside the television industry. A few may be appointed to manage a department concerned with programme-making, but most are appointed in areas such as Computing, Legal, Personnel, Training, Accounting or Administration. These functions are found in most organisations and consequently there is considerable movement between industries. Such Managers usually have the appropriate professional qualification for the area in which they work.

A small number of Managers are recruited from time to time as trainees by individual ITV companies. They may be given a broad training in a variety of departments in order to gain an overall understanding of the business, and then they may specialise in a particular area. There are no regular intakes of management trainees, and the numbers recruited are very small. Most companies will advertise in the press when they have vacancies. Successful candidates may typically have a degree or professional management qualification, plus a short period of experience in another industry. There are, however, occasional opportunities for internal applicants from any area to apply for posts as trainee Managers.

There is no recruitment profile given here for the post of Manager, since there is no one background or qualification which is essential. Professional management qualifications such as an MBA (Master of Business Administration) and DMS (Diploma in Management Studies) are becoming increasingly recognised as important in an industry as competitive and commercial as television. The key requirements are the ability to plan, organise, motivate staff and control resources. Managers are expected to work long hours, and it is potentially a seven-day-a-week job.

Training for new Managers is tailored according to their needs, and may include formal off-the-job training to acquaint them with the industry, or provide them with specific management skills.

Management in television offers variety and challenge in a demanding environment, and the prospect of a first-class career.

PERFORMERS

Actors, dancers, musicians, and stunt performers are always employed on contracts for fixed terms. There is no such thing as a performer on the staff of a television company. Even those engaged on long-running series or serials will be contracted for periods which will rarely exceed twelve months, although there may be options to extend the initial period.

Actors and Dancers

The majority of actors (and other performers) regard television as one of the media in which they expect to work, and their experience will include acting on the stage, in films, and on radio. The job is allegedly glamorous and for this reason, there are always far more people available to act and dance than opportunities for doing so. It is estimated that at least two-thirds of actors are unemployed at any one time. The actors' trade union, Equity, has consequently sought to restrict the numbers by ensuring that offers of employment in television, the West End theatre, major subsidised theatre companies, feature films, commercials, etc., are made only to actors and dancers with previous experience, who are Equity members.

Details of the kind of work which is available to non-Equity members can be obtained from Equity, at 8 Harley Street, London W1N 2AB.

There are a number of dance and drama schools throughout the country which offer training, but their standard varies. The best will give a thorough training in performance. It is also possible to achieve training through experience in a repertory company, but this is becoming more difficult with the gradual decline of the repertory movement.

It is sometimes believed that one way to obtain experience as an actor is by working as a 'walk-on'. This is a performer who is not required to give an individual characterisation, but usually forms part of a crowd scene. In fact, there is an agreement in the major cities of the UK, similar to the one for the main performers. This limits 'walk-on' work, in the first instance, to members of Equity. It is doubtful in any case whether this work offers any real opportunity to the aspiring actor.

Most actors and dancers obtain work by signing a contract with a theatrical agent. A good agent will actively seek the right kind of work for the individual performer, and will have extensive contacts and experience in television, film and theatre. When performers are required for a television programme, the Producer or Casting Direc-tor may approach someone they have seen in another production and feel is appropriate for the part. They may also approach agents to recommend suitable people, and search through casting directories such as *Spotlight*.

Musicians

The great majority of musicians are casually employed for a particular programme or series. They are usually engaged for a recording session or performance for periods of about three hours, and should generally be members of the Musicians' Union. Some companies employ a Head of Music and a Musical Director on the staff. They are responsible for contracting composers, arrangers, and musicians and are always highly experienced. They must have a wide musical knowledge and be capable of advising Producers on all matters concerning the musical content of a programme whether it is pop music or a symphony orchestra.

Stunt Performers

Employment as a Stunt Performer in television is normally restricted by Equity to members of the Equity Register of Stunt Performers and Arrangers. To register as a Stunt Performer it is usually necessary to have experience as a professional actor. Stunt Performers are regularly called upon to double for actors, and to act in their own right.

Certain qualifications are recommended, and it is advisable to check with Equity for the latest details. There are a number of categories of stunt and there are appropriate qualifications for each. The 'Fighting' category, for example, includes qualifications in fencing, judo, other martial arts, wrestling and boxing. The 'Falling' category includes qualifications in trampolining, diving and parachuting, while the 'Agility and Strength' category includes qualifications in gymnastics, weight training and ballet and athletic dance.

The world of acting, dancing, and making music is highly competitive and there are many hurdles to be overcome. It is, however, a highly rewarding career for those with talent who are fortunate enough to be in the right place at the right time.

PRODUCER

Many people do not realise that the role of the Producer is different from that of the Director. Of those who do, a good many would be hard-pressed to explain what the differences are. This is not surprising. Often in television the same person wears both hats and assumes the title of Producer/Director. So where does the Producer fit into the scheme of things?

The Producer heads up the production team that comes together to make a programme or a series of programmes. If you think of a television programme as being like a manufactured product, it is the Producer's responsibility to ensure that the production team produces the goods on time, at the budgeted price and quality. To do this, the Producer calls on a wide range of specialist and service departments who will advise on how best to carry out the various stages of the production process. Some of this advice will be conflicting and there will also be uncontrollable extraneous factors to take into account. In the end the vital decisions are the Producer's alone. There is a budget to work to, and the Producer must decide how best to spend the money to achieve the objectives in terms of delivery time, quality and price.

Usually the Producer will be either the originator of the idea on which the programme is based or will have made a substantial contribution to the development of that idea. Sometimes, Producers are lobbied with ideas so it is their job to select from a number of programme possibilities. The Producer will arrange meetings of the production team or with individuals to discuss the development and realisation of the idea and then recruit or commission any script writers.

In television it is the rule rather than the exception for the Producer to make a significant creative contribution to the production. The actual shooting of the programme and the direction of the performers and technical crews are the Director's responsibility. The Producer may, however, take a hand in selecting the actors or in choosing the locations for a play, and will certainly try to pick the best people for the production team. Ultimately, it is the Producer's responsibility to ensure that people and objects are where they should be, when they should be, for the shoot.

At all times, the Producer's desk is where the buck stops. If bad weather delays outdoor shooting, or the leading actor falls ill, or even (as actually happened with one major ITV drama series) the studio and all the scenery are destroyed by fire, the Producer still has to get the programme made on time. Only in very exceptional circumstances is an over-spend on the programme budget allowed.

We have tended to take as our example someone working in drama, but in fact every programme needs a Producer, who must possess experience and skills appropriate to the type of programme being made. Thus, to succeed in producing light entertainment, a sure instinct for popular appeal and an eye for new talent are essential. In current affairs, sound political and editorial judgement are paramount while in sport the skills of commercial negotiation are almost as important as knowledge of the subject matter. Producers generally are intelligent, articulate people with a high level of social and administrative skills.

As with the Director's job, there is no typical career path leading to the job of Producer. Many are ex-Directors or ex-Researchers but just about any job which requires a comprehensive understanding of production techniques can equip one for promotion to Producer.

PRODUCTION ASSISTANT

The job of the Production Assistant (PA) is changing, like most jobs in television, as new technology removes the need for certain tasks to be carried out manually and simplifies others. The changes which are occurring vary from company to company and depend very much on the type of programme being made. The duties described here, however, give a broad idea of the range of tasks that may be carried out.

PAs are an essential part of the production team and provide both the organisational and secretarial services for the Programme Director. They are in essence the Director's personal assistant and are generally assigned to a particular programme from the start to the finish. Some may, however, move on to another programme before the first is completed. In some companies, mainly the larger ones, PAs may specialise in a particular type of programme, such as sport. More experienced PAs may specialise in drama, but in many companies PAs work on all types of programme produced by that company, and all will have their share of live programmes, such as local news, at some time.

The Programme Director's office is the focal point to which everyone refers during the making of the programme, and it is the PA's responsibility to coordinate all the various activities.

During the planning stage of a recorded programme, PAs accompany the Programme Director to the many meetings with the production team (design, sound, cameras, lighting, etc.), and with the artists. They make copious notes of all decisions and make sure that the required action is taken. This inevitably involves a considerable

Production Assistant checking continuity with an actor (YTV)

amount of routine office work such as booking rehearsal rooms, technical equipment, hotels and catering facilities.

Next comes the rehearsal period and, since the time spent in the studio is very expensive, most rehearsals are held outside the studios. During rehearsals for a drama series or a situation comedy, for example, the Programme Director may decide to change the script and other details several times. The PA must note any changes, and type and retype the script until the Programme Director is happy. This work is often done under pressure, so fast and accurate typing skills are essential.

The final 'camera' rehearsal takes place in the studio and is a final run-through for all the artists and the production team. From this point, through to the end of the programme, PAs sit alongside the Programme Director in the control room. One of their roles is to call out instructions to the camera operators via the talkback system so that each of them is reminded when they will be 'on air'. Another critical function is to keep an accurate check on the timing of each part of the programme since it will usually be recorded out of sequence.

A stop-watch is used to ensure that the finished programme will be exactly the right length. At the same time PAs must make notes of any further decisions taken by the Director, so they must be able to cope with several tasks at once. The ability to concentrate and think quickly is essential and this is not an easy task at the end of a long day's work in a somewhat claustrophobic control room which has no daylight.

On completion of recording, most programmes require some 'post production' work, i.e. editing and the addition of further sound tracks.

If the technicians working on these jobs require any information, it is to the PA that they turn. The picture editor in particular relies heavily on the notes supplied by the PA.

Some productions may involve assignments on location and, once again, the PA will keep accurate notes of the proceedings. These may include continuity notes to ensure that the visual flow from one scene to the next is correct. A high degree of perception is needed for this task. Location assignments may also require the PA to be away from home for periods of time.

So far, we have referred mainly to recorded programmes, but PAs also spend a significant amount of time working on live programmes in the studio, such as local news. This is particularly so in the first year or two of a PA's career. Rehearsals for a local news programme are restricted to a quick run-through of the various items about an hour before transmission, and there is no post-production work to be done. The duties of the PA in the control room are very similar whether the programme is live or recorded. The main difference is that with live programmes there is no room for error and the situation can therefore be very stressful at times. Since local news programmes tend to follow a set format, a good PA will soon learn to anticipate any problems that might arise, and take appropriate action.

The PA's job is far from a nine-to-five one, and irregular hours frequently impinge on private lives both on weekdays and at weekends. The work is demanding and often means working under pressure in inconvenient and uncomfortable circumstances. Taking notes on location when your fingers are frozen and you are wrapped in several layers of woollens is not easy!

PAs must be able to relate to all types of people in the course of their work, including performers, VIPs, members of the public, and other members of the production team. Some of these people will be nervous or under pressure, and may not be in a cooperative frame of mind.

Recruitment to the post of Production Assistant is usually at trainee level. Competition for posts is very tough and vacancies are often filled by staff already working for the television companies. Many of these are secretaries and some are in administrative posts. The demands of the job mean that the standard of applicants in terms of personal qualities and educational qualifications must be high.

The work is, however, rewarding and with increased experience PAs may go on to work on even more prestigious and difficult programmes. This is a job which, more than almost any other, provides an all-round knowledge of the world of television.

Typical Recruitment Profile

Trainee Production Assistant

	Essential	Desirable
Physical		
Age	20+	22+
Health	Good physical stamina	
Eyesight	Normal eyesight and colour vision	
Hearing	Good	
Speech	Clear speech	
Education and Training	Broad general education to GCSE grade A to C standard. Subjects to include English. Fast, accurate typing (about 40 wpm) and shorthand (about 100 wpm) or speed-writing. General secretarial skills	'A' levels or degree (subject not important)

	Essential	*Desirable*
Experience	Shorthand and typing	Previous secretarial experience and/or experience of organising. A working knowledge of computers or word processors
Interests	Varied interests including television, current affairs, etc.	Films, theatre, music. Ability to read music
Personality	Ability to work accurately and calmly under pressure, assess priorities, organise and think quickly in an emergency. Sensitivity, tact and confidence in dealing with contacts inside and outside the company. Good powers of observation, and an aptitude for mental arithmetic. Able to listen and interpret instructions accurately. Able to work as a member of a team	
Availability	Able to work long and irregular hours at any location	

PROGRAMME DIRECTOR

A television programme of any kind invariably starts life as a collection of ideas, probably embodied in a script. The Programme Director's job is to take those ideas (to which he/she may well have contributed already) and to translate them into the sequence of pictures and sound ultimately seen by the viewer.

The translation requires the skills of a great many people. It is the Director who plans and controls the activities of those people and to a large degree it is the Director's skill at motivating them to give of their best that determines the excellence of the finished programme. In order to do this successfully the Director needs a thorough appreciation of everybody's job and the technical or artistic limits within which they are constrained to work. For instance, the IBA sets rigid technical standards for the quality of the pictures and sound in a programme. If the Director wants to create a particular visual 'mood' effect such as the warm glow of a fire in a darkened room, the engineers and lighting director can easily produce it. But the actors will not experience that mood easily because even the darkest parts of the set will need to be quite brightly lit in order to achieve the correct technical balance. An experienced Director will know this and will plan accordingly.

At its most sophisticated level the job of the Director is to create say, a television drama, shot by shot, coaxing the utmost from actors, camera operators, sound, stage crew and all the supporting technicians. When shooting is complete, there follows a long period of post-production editing and sound dubbing which the Director must supervise.

At a more basic level, for instance on a news magazine programme, the Director follows a running order of items, selecting pictures from the shots offered by the camera operators and relaying 'speed-up', 'slow-down' instructions to the presenters.

Both these examples present their own challenge: the first, that of achieving artistic excellence; the second, that of converting chaos into order. Both demand a high degree of leadership and the ability to get on with all types of people.

Clearly there are Directors and Directors. Not all are equipped to work on a prestigious network drama. By mid-career most Directors will have found their level and/or their forte. Drama is not the only thing that calls for special directorial skills: light entertainment, sport, and music are all areas where reputations can be made. Every company needs a number of competent general Directors who, though they may not be truly gifted, can turn out a good programme in any area. Many companies require Directors to have equal facility with both video and film production although, as video cameras get smaller and lighter, the differences of technique between film and video are diminishing rapidly.

Some Directors are also Producers, but the combination of the two roles can make heavy demands on their time.

There is no clear-cut path leading to the job of Director. A very small number of theatre Directors make the transition to television by direct entry. In the main though, trainee Director posts are filled from within the industry by people with substantial experience of production. Traditionally the most popular sources of directorial talent have been Floor Managers, Camera Operators and Researchers, but this list is by no means exclusive. Just about any job in television production can be the stepping-off point for a career in direction.

The competition for trainee posts is intense in the extreme. For those who make it to the top, the rewards, both in job satisfaction and income, are substantial.

Trainee Programme Director

No recruitment profile is given for this job, as it is impossible to describe a 'typical' Director. Candidates will in general be already employed in television, in jobs described in other sections of this book.

PROGRAMME RESEARCHER

The job of Programme Researcher in television is a much more broadly based and demanding job than the title suggests. It has few parallels with the kind of academic research familiar to undergraduates. The majority of Researchers work in the area of current affairs programming where their job is journalistic, similar in many ways to that of a newspaper reporter. Every ITV company has its own evening news magazine programme. This is where most young Researchers cut their teeth. They are expected to contribute ideas for the programme and, once the format is decided, to prepare the material for a particular item. This will include identifying and interviewing contacts, obtaining appropriate film or video tape material (usually by going out with a small team to record, but sometimes also from library sources), getting interviewees into the studios and writing a script or 'treatment' for the Presenter. The range of tasks undertaken by Researchers at this level is virtually limitless, from finding participants for a quiz show to tracking down an obscure picture or piece of film of a long-forgotten event that is suddenly topical again. Moving beyond the very general duties of this kind, Researchers tend to specialise. In the bigger companies the goal of many is to work on a prestigious network current affairs programme.

Here the preparation time is generally measured in weeks or days rather than hours, and much of the work is best described as investigative journalism. Researchers work in teams with the Producer and Programme Director, often travelling extensively with a production unit and sometimes exposed to serious personal risk.

Outside the broad field of current affairs are the more specialised Researcher roles, most of which call for a high level of specialist knowledge and skills. These include research for programmes in such areas as natural history, science, anthropology, music, and light entertainment. Drama research, although an identifiable job in television, does not employ significant numbers of people. Much of the research needed for a drama production is done by the Writer, the Designer, and the Make-up and Wardrobe specialists. A significant proportion of Researchers are employed on children's and educational programmes.

The majority of Researchers are graduates, although few companies recruit by direct graduate entry. It is normally necessary to have had some postgraduation experience in a media-related area such as newspapers, radio or television. As a body, Researchers tend to be forceful, extravert, creative and intensely competitive. Their natural career progression is to Producer and less commonly, to Programme Director. They can also become Presenters and even occasionally, News Readers.

It is normal practice in the industry for Researchers to be engaged, at least initially, on renewable fixed-term contracts. There is significant movement of Researchers between companies, particularly with the true specialists who can work only for a company that is making programmes in their subject area. Such people are normally engaged on a 'run-of-series' contract.

Typical Recruitment Profile

Trainee Programme Researcher (General)

	Essential	Desirable
Physical		
Age	21 +	23 +
Health	Good stamina	
Education and Training	Broad, general education, usually to degree standard	Good class of degree. A degree in the appropriate subject is highly desirable for specialised researcher posts
Experience		Media-related, in newspapers, radio or TV. Knowledge of reference sources. A background in journalism is highly desirable for posts in current affairs
Interests	Broad, embracing the arts, sciences, current affairs, television	
Personality	Articulate, inquisitive, resourceful, creative, resilient, extravert and competitive but a good team	

	Essential	*Desirable*
	member. Able to relate easily to all types of people	
Availability	Able to work irregular hours, anywhere, anytime	

SALES AND MARKETING

The Sales Department is a vital part of every ITV programme company. Nearly all of a company's income is derived from the sale of advertising time, and if the Sales Department fails to sell that time, there is no money to make programmes and pay salaries.

Although television is a very effective medium for advertising, the task of the sales staff is not simply to wait for the bookings to come in. They are competing not only with the press and radio for advertisers, but also with the other ITV companies and increasingly with cable and satellite television stations. Staff must therefore have a very positive approach towards existing television advertisers and towards getting new business.

The ITV companies have sales offices in London, and in some other parts of the country where there is likely to be plenty of business. (Addresses of sales offices can be found in the section entitled 'How to apply for jobs in ITV'.)

Since advertisers wish to reach particular members of the population, perhaps in certain parts of the country, all advertising time is sold on a regional basis by the local ITV station. The exception is TV-am whose sales are handled nationally. Since both ITN and Channel 4 have national coverage, their sales are handled locally by the ITV regional companies and they do not employ their own sales staff. All of the ITV regional companies and TV-am employ sales staff.

As soon as the ITV network produces its schedule of programmes for the next quarter of the year, the sales department can begin to sell the advertising spots or 'airtime'. There are considerable differences between the various sales jobs in different companies but the following general descriptions should be a useful guide.

Sales Coordinator

In some companies, these tasks are carried out by Sales Negotiators or Sales Executives, but for the sake of simplicity we will refer to Sales Coordinators.

The aim of the Sales Coordinators is to obtain maximum revenue through the sale of airtime and to ensure that every commercial break is filled with advertisements. They receive bookings over the telephone from advertising agencies or direct from clients, and refer to the computer to see what airtime is available for sale.

In addition, they must take positive action themselves to fill empty spaces, and use their knowledge of airtime availability and advertising campaigns to decide who to contact in the hope of persuading them to buy.

They will discuss overall campaign details and requirements. An agency might for example handle a client that sells holidays and wishes to mount a campaign on television from Christmas to March. The Sales Coordinator must be aware of any such campaign and be informed as to the size of budget to be spent and the type of person at whom the advertisement is aimed (perhaps in this case women in the middle-income bracket). They will then be able to suggest the best programmes or commercial breaks in which to place the advertisement across the campaign period, subject to availability. The Sales Coordinator will book the airtime on the computer and, as time goes on, handle any booking changes that may be required.

Sales Coordinators are office-based for most of the time and constantly use a telephone and a Visual Display Unit (VDU). They must be highly articulate, outgoing and pleasantly persuasive with a good telephone manner. Numeracy is also a vital attribute since it is often necessary to calculate the different rates for advertisements quickly and accurately. A knowledge of ITV programmes, audience research data and current affairs is also needed.

A smart appearance is important as from time to time, Sales Coordinators will meet clients and representatives from agencies. They generally work in teams and will be responsible for a particular group of clients and/or agencies.

Most trainees are appointed in this area, particularly if they have no previous television advertising sales experience. Selling airtime is an ever-changing and complex business, as well as being highly competitive. Each company therefore operates its own training scheme, the length of which may vary.

Marketing Executive (or Senior Sales Executive, or Sales Executive)

These staff are responsible for attracting new business to the company and for providing support for existing clients and agencies. They spend less time on the telephone and much more time visiting customers. Their task is to promote the concept of television as an advertising medium and in particular to promote their own company. They may give audio-visual presentations which provide statistics for potential customers to show why they should buy airtime, as well as giving details of the cost of a campaign. They may also discuss how a campaign might be organised. Marketing Executives may have to travel extensively around the UK, giving promotional talks and generally looking for new business.

The personal qualities that make a good Marketing Executive are similar to those of a good Coordinator.

Marketing Executives usually have supervisory responsibility for Sales Coordinators.

Market Research Executive (or Sales Researcher)

Market Research is an essential support function for the Sales Department. The job of the Market Researcher is to collect and interpret market research and media research data for a wide variety of projects. The Sales Executives may, for example, wish to know how effective television advertising has been in selling breakfast cereals in a certain part of the country so that they can persuade a potential new client to mount a campaign.

The Market Researcher extracts the data, interprets it and if required prepares it for presentation to the agency and/or client. This could entail the commissioning of original market research, which is placed with external research agencies. The Market Researchers may, however, devise the questionnaires to be used, and tell the research agency what is needed.

In some companies, data is collected for Programme Producers on the type of people watching various programmes, while some also provide statistics for management to help forecast revenue and growth trends.

Good Market Researchers must be highly numerate in order to interpret data, and sometimes to develop computer models. They may well have to spend long periods working with a VDU. The ability to relate easily to staff in other departments and understand their requirements is essential. In some companies, Market Researchers may also be called upon to make formal presentations of their findings to customers, and all should be able to write clear, concise reports.

Trainees may be appointed as Research Assistants, helping the Market Research Executives in all aspects of their work.

Traffic (Make-up Clerk)

One of the chief functions of the Traffic Department is to monitor the 'make-up' of the commercial breaks in the few days (or perhaps weeks) before transmission. As the transmission time draws near, traffic staff handle the administration of any last minute bookings.

At this stage it is important to check that the strict rules laid down on advertising by the IBA (Independent Broadcasting Authority) are being met, although of course staff must always be on the look-out for problems. An actor who is appearing in a programme, for example, must not appear in an advertisement being shown during that programme, and certain advertisements must not be shown within a given time of a children's programme. A good knowledge of ITV programmes is therefore essential. Equally important is a keen interest in current affairs; for example, if there has been an air crash, certain airline advertisements might be in bad taste and the traffic staff would be instructed to ensure that they were withdrawn.

As transmission time draws even closer, traffic staff number the advertisements in the best order for viewing. It makes life easier for the transmission staff if all the advertisements which are on video tape are together, and all those on film are together.

In some companies traffic staff may also arrange for the collection and delivery of advertisements. They may carry out other duties such as updating the computer on changes to television programmes and entering other data relevant to the sales staff.

Traffic staff spend a great deal of time on the telephone and they must be prepared to work with VDUs. They must be thorough and conscientious in checking the commercial breaks before transmission.

Trainees are generally recruited into sales coordination, research or traffic. In some cases, however, applicants with good 'A' levels, or a degree or equivalent, may be recruited into higher grades (e.g. Trainee Marketing Executive). Some experience of brand selling or working in an advertising agency can be useful.

Opportunities exist for promotion through the various grades, or staff may move, say, from traffic to sales coordination.

Television Commercials

The ITV companies are responsible only for the transmission of commercials, not for making them. Advertising agencies are responsible for the creative aspects of advertising campaigns, and further details can be found in a useful publication *Getting into Advertising*, which is available from:
The Advertising Association, Abford House
15 Wilton Road
LONDON SW1V 1NJ
Tel: 01 828 2771

Typical Recruitment Profile

Trainee Sales Staff

	Essential	Desirable
Physical		
Age	18 + (21 + for some marketing posts)	
Appearance	Clean and smart. Acceptable to all types of people	

	Essential	*Desirable*
Speech	Good clear telephone voice	
Eyesight	Normal (able to work with VDU)	
Education and Training	Two or more GCE 'A' levels. Subjects not important, but Marketing, Business Studies, Computer Studies, Maths, Statistics or Economics are appropriate	A degree. The subject is not important, but the same subjects are appropriate as for 'A' levels
Experience		Telephone sales, brand selling or working with an advertising agency
Interests	A broad range of interests including current affairs and social activities. A knowledge of television programmes	
Personality	Outgoing and friendly, articulate, numerate. Able to work as a member of a team, and able to relate to all types of people	Leadership
Availability	Normal working hours in junior posts, but increasingly irregular hours with promotion to higher grades	

SCRIPT EDITOR

There are relatively few people employed as Script Editors in the television industry, and yet the job which they do is both essential and rewarding.

They are nearly all employed in the drama department, working directly for the Producer of a particular programme. Their role is to encourage Writers to develop and express their ideas to the best of their abilities, while ensuring that the Producer's programme needs are met. Experienced Script Editors are usually responsible for commissioning Writers, either alone or with the Producer. Many of these Writers will have a well-established reputation, but Script Editors are always on the look-out for fresh talent and new ideas. Their search will often take them to fringe theatres, film shows, and to libraries and festivals.

The kind of Writer that the Script Editor will look for to write a script for a single play will be very different from the kind of Writer needed for a serial. In the former, a number of Writers are needed with a mixture of individual ideas perhaps linked by a common theme. They will each contribute a play which will form part of a series. In the latter case, the serial will be originated by one Writer, but others will be needed to carry on in the same style over the months and years to follow.

Once the Writer has been commissioned, the Script Editor may be concerned with almost anything to do with the script. There is of course a great deal of administration and correspondence to be done in order to ensure that the script is ready on time, but other duties will depend largely on the way in which the Producer prefers to work.

Some Script Editors carry out research, for example, into historical facts, but this is uncommon in ITV. Some will prepare the 'storyline', liaising with the Producer. This is a rough outline of the story which the Writer then brings alive. Others may carry out 'rewrites' of certain parts of the script if this work is unavoidable and cannot be done for some reason by the Writer.

Script Editors who are working on serials need an excellent memory. It is their responsibility to advise the Writers on individual characters, their relationships, and who has done what in previous episodes. There would be a considerable outcry from the viewing public if an actor was asked to do something which was out of character, or if an incident was repeated. The Script Editor is therefore responsible for guiding the Writers, moving the story along and for keeping the script within transmission time and budget.

The job requires a very high degree of diplomacy. Everyone can be sensitive to criticism of their work, and Writers are no exception. The Script Editor needs to exercise considerable tact and discretion (as well as forcefulness!) when advising a Writer how to tighten a script, or advising that a particular scene or incident will not work dramatically or visually. The role is that of a mentor, and the task is to produce a script that has the highest possible dramatic quality. The skill is to be constructive rather than destructive and to encourage rather than stifle creativity.

It is impossible to describe a typical career path into this job since Script Editors come from a variety of sources. A good academic background can help to develop the strong critical faculties that are needed, and a degree in Drama, English or a subject related to Communications is therefore an advantage. Experience as a Script Reader for a fringe theatre is also a useful way of developing a sense of judgement and confidence. Script Readers are generally paid minimally, but are employed by most theatres to read and prepare reports on the many unsolicited scripts they receive from aspiring writers. This is a useful way of learning to discriminate between work which is promising and work which is merely run-of-the-mill, and developing the confidence to ensure that you are not going to reject a budding Shakespeare. Script reading for a film or television company is also useful experience, as is a background in journalism. It is unlikely that anyone without a deep interest in drama of all kinds and in communication would be attracted to this kind of work.

Other essential qualities are self-motivation, patience, a willingness to work hard and variable hours, an unquenchable interest in how writers write, and the ability to work well under sustained pressure.

No 'Typical Recruitment Profile' is given for this job.

SECRETARY AND CLERK

There is a very wide range of secretarial and clerical posts in the ITV companies and it is impossible to describe here all the openings that may be available. Secretaries and Clerks are employed in almost every department and provide an essential support service in production areas, newsrooms, technical areas, Sales, Personnel, Accounts, etc.

Clerks

Junior Clerks are employed in jobs such as handling post, duplicating and printing. More Senior Clerks may be employed in posts which carry a fair degree of responsibility such as the booking of British Telecom lines for sending programmes from one station to another, or monitoring the booking of advertisements into commercial breaks. The range of jobs is endless. For those with an aptitude for figures, there are openings in Wages and Accounts departments, and for those who enjoy working with people, there may be openings in Personnel or production areas.

Some companies ask for a minimum of five GCSEs (grades A to C) from applicants for clerical posts and although others do not make this stipulation, it is important to remember that competition for vacancies is tough, and qualifications are an advantage. An ability to type and a good telephone manner are also useful in some jobs.

All of the ITV companies look for applicants with good basic common sense, initiative, the ability to work as a member of a team, an interest in the work, and a conscientious attitude. Manual dexterity is very useful as many jobs involve the operation of a computer keyboard.

Applicants for junior clerical posts are generally accepted from the age of 18.

Junior Secretaries

Applicants for junior secretarial posts should have a typing speed of about 50 words per minute, preferably with an appropriate qualification such as RSA II or LCC Intermediate. Previous secretarial experience is not essential, but can be useful. Tasks include filing, dealing with telephone calls, maintaining catalogues, typing, shorthand, audio typing on occasions, and general office duties. Junior Secretaries may work alone, or in a group with other Secretaries and Clerks depending on the nature of the department.

GCSEs grades A to C or equivalent qualifications are an advantage, and some companies ask for a minimum of five passes. The personal qualities which have been mentioned under the heading of clerical posts are equally important for applicants for all secretarial jobs, both junior and senior.

▥ *Secretarial*

Secretaries may work for one person, or for several. They may work alone or in groups. Applicants for more senior positions should be able to organise themselves and the office, use their initiative, and work under considerable pressure from time to time. They should also be able to relate easily to other people, both on the telephone and face to face.

Good educational qualifications are essential and most companies prefer applicants to have five GCSEs (grades A to C) or equivalent, preferably including English Language. Applicants with GCE 'A' levels and a degree or Business Studies qualifications are also welcome, particularly for more senior positions.

Previous secretarial experience is useful for all jobs, and essential for many. Applicants should have a typing speed of 50 words per minute and a shorthand speed of at least 90 words a minute. Experience with word processors and computers is an advantage.

Clerical and secretarial posts provide a rewarding career for many people, lasting the whole of their working lives. Others see such jobs as a stepping stone, perhaps into the programme-making side of television or into administrative and management roles. Opportunities for movement into these areas are highly unpredictable and no guarantees can be given.

Nearly all vacancies for Trainee Production Assistants are filled by Secretaries from within the companies. In addition, some Secretaries work very closely with the production team. An example is the Producer's Secretary, Production Secretary or Programme Secretary (not employed by every company) who carries out many of the typing functions of the Production Assistant. Programme Secretaries have to become proficient in the transcribing of cassettes containing the sound from filming, so that the film and dialogue can be edited down together.

All staff are expected to spend a reasonable time in their first job before applying for other posts in the industry.

No recruitment profile is given for clerical and secretarial posts as jobs vary so greatly. In the main, the requirements are no different from those for similar jobs in any industry.

SET DESIGNER

The job of the Designer is one of the most creative in television, and yet it is one which the viewer largely takes for granted. Almost every programme from the studio interview to the prestige drama has a Designer whose job is to create the right sense of mood, style, period and place for the programme. The Designer creates the setting within which the action takes place.

Television is a creator of illusions, and so the work of the Designer is not made to last. A solid-looking interior of a house will be made from thin wood, and a dense jungle will be created with a few potted plants. As soon as the recording is finished, the set is dismantled, and usually destroyed.

The work of the television Designer is more complex than that of the Theatre Set Designer. A theatre audience views the set from one angle and from a distance, and sees the set as a whole. A television audience sees the set in close-up, in small sections and from many angles, each of which must look equally realistic and cohesive. The job therefore requires meticulous planning, particularly for dramas and light entertainment programmes where the cameras may be required to move around a great deal. Changes in technology are leading to even greater clarity in the picture on the domestic receiver, and this demands a sharpening of all the traditional skills such as attention to finishes, accuracy of detailing, and set dressing.

Designers develop their own way of working. Once they have been allocated to a particular programme, for example a drama, they will usually begin by reading the script and translating it into visual terms by drawing sketches. This demands creative interpretation of the writer's intentions. The Designer works closely with the Programme Director. Both will have their own ideas on what should be done, but Designers are usually allowed a considerable amount of freedom. They must of course work within a strict budget which is allocated to the programme. They will often draw a 'storyboard', which is rather like a strip cartoon, to show the progression of the action and help in

Set Designer making a scale model of a set (YTV)

the creation of sets. The next step is usually the production of simplified architectural drawings of the sets from which costing estimates can be made by another department. This may be done manually but computer-aided design (CAD) systems enable the Designer to plan accurate perspectives of shots that will be available to the Director on the recording day. Ground plans of the studio are also made to ensure that the sets will allow room for cameras and microphone booms to be positioned and moved, for the lighting to be correctly positioned, and for actors to move around.

There is obviously a great deal of liaison with people in other departments such as the Programme Director, Chief Costume Designer, Lighting Director, Sound Supervisor, etc., and several meetings are held to make sure that all their interests are met. The Designer may produce scale models of the proposed sets for these meetings, and these are of considerable help in ensuring, for example, that a camera can be moved from one position to another without its cables getting in the way.

Once the design of the set has finally been agreed, the construction workshop starts to assemble the scenery, or the basic parts of the set are hired from a specialist firm. Other items are made, hired or bought. Some parts of the set may be 'topped up' or other parts such as ceilings may be added electronically to the picture after the programme has been recorded. The Designer achieves this by using complex computer graphics equipment.

Designers are responsible not only for the scenery, but for everything else that is on the set such as curtains, furniture, books and carpets. Considerable research is often needed to make sure that the whole effect is authentic, and Designers may spend many hours researching in libraries, museums, etc., before preparing a 'prop list' of required items which they will then obtain from a wide range of sources.

If part of the programme is to be shot away from the studio, the Designer will usually help to find suitable locations, and will take steps to hide any unwanted views. An authentic period cottage, for example, may have a bus-stop outside it which does not fit into the intended period of the programme, and will have to be disguised, or even temporarily removed.

During studio rehearsals, the Designer will supervise the erection of the sets and the 'dressing' of the sets (adding all the extra details). Minor adjustments may also have to be made. When the recording takes place the greater part of the Designer's work is already done and they will probably already have started work on other programmes.

Recruitment is usually as an Assistant Designer. Assistants normally help in the preparation of drawings, models, etc., and in any research that has to be done. They gradually take on more responsibility until a vacancy occurs for a Designer.

Typical Recruitment Profile

Trainee Assistant Set Designer

	Essential	Desirable
Physical		
Age	21 +	
Eyesight	Good colour vision	

	Essential	Desirable
Speech	Able to communicate clearly	
Education and Training	BA in Interior Design or BA in Art and Design or Architecture Degree or BA in Stage Design	
Experience	Draughting	Commercial art studio, interior design, theatre design, architecture
Interests	Three-dimensional design, architectural styles, contemporary design, period furniture, fashion trends	Theatre, television, films
Personality	Ability to communicate ideas. Ability to work as a team member. Highly creative. Able to work under pressure at times	
Availability	Willing to travel and work unsocial hours when required	

SOUND TECHNICIAN

In a medium which is dominated by the picture, the contribution of sound is often underestimated by the viewer and yet programmes without theme music, background music and sound effects, let alone dialogue, would be unthinkable.

There has been a gradual decline over the years in the number of television programmes which are made on film. The great majority are made on tape, and the emphasis here is on taped productions. The operational and artistic skills which are required by the sound specialist are, however, very similar regardless of whether film or tape is used.

Studio Sound

The early years of a Sound Technician's career are normally spent in the studio gaining a thorough grounding in studio sound operations on the studio floor. A period of familiarisation may also be spent in the sound control room.

One of the first skills to be grasped is that of operating the boom. The boom microphone is mounted on the end of a telescopic arm which in turn is mounted on a dolly. This is moved around the studio to follow the action. Good hand–eye coordination is necessary, together with an understanding of camera angles and studio lighting. Incorrect positioning of the boom will not only produce poor perspective sound, but may cast unwanted shadows from the lights across the picture. This can result in a costly retake of a scene.

Sound Technicians will learn the characteristics and uses of the various other types of microphone in the studio, and how to place them correctly. They will also be responsible for ensuring that the communication systems such as 'talkback' between the studio and control room, as well as loud-speakers, etc., are working.

Sound Mixer

A variety of duties are carried out in the Sound Control Room including the monitoring and adjusting of sound levels during recording. The staff who work here are generally more experienced, and include the Sound Mixer, who is normally a Supervisor. Sound Mixers are responsible for the crew working for them, and for the operation of the audio mixer console or 'sound desk'. This consists of a vast array of buttons and switches, each of which is concerned with an incoming sound source. There may be typically 24 separate sound sources or as many as 56 which the Sound Mixer can fade up or down, mix and balance to produce the desired output for the programme. The Sound Mixer will also correct deficiencies in the original sound as far as possible.

Grams Operator

One of the sources feeding into the sound desk will be from the Grams Operator, who has a library of sound effects and music. These are held either on tape or on compact disc. On cue from the Director, the Grams Operator will play the appropriate material into the programme, and the Sound Mixer will make any necessary adjustments.

During live transmissions such as news and current affairs, Grams Operators may be required to work on their own initiative selecting appropriate music and effects for the programme. Creativity and good aural perception are essential for this. The Grams Operator must also be able to edit tape under the kind of pressure caused by tight deadlines.

Post-Production Sound

Once the programme has been recorded, the finishing touches are put to the sound in 'post-production'. The picture will have been edited and re-assembled into its final form, and if left untouched, there would probably be jumps and discrepancies in the accompanying sound tracks. The technicians in post-production sound, or dubbing, re-assemble and even re-record the sound track to fit the picture. They may also add special sound effects which were not included during recording, as well as the opening and closing theme tunes for the programme. Occasionally there may be an unwanted sound such as aircraft on the original sound track. This must be removed or disguised so that the viewer is unaware of what has happened.

Attention to fine detail is essential. It may be necessary to match the voice recordings of one artist made in, say, 23 different locations, each of which has different acoustics.

Experienced Sound Technicians with proven artistic and creative abilities are normally selected for these positions.

Outside Broadcasts

All of the duties which have been described so far are equally important on outside broadcasts, although the equipment used may be slightly different, and there are some different challenges to be faced. Whereas in the studio, much of the equipment is fixed, every facility to be found on an OB (outside broadcast) has to be set up and made to work by the crew on the day, and then stripped out after the event. This is often done in darkness and rain.

Sound Recordist on training exercise (Thames TV)

Sound Technicians tend to have a variety of responsibilities for sound on an outside broadcast, and specialist knowledge is needed, for example, to cope with the wind blowing against microphones at a sporting event, or gathering sound effects over wide areas and long distances. The OB sound crew is also responsible for communications, which are often extremely complicated.

Maintenance

The responsibility for the maintenance and technical alignment of the sound equipment varies from company to company. In some cases, these tasks are carried out by engineers in a separate maintenance section, while in others, it is the responsibility of the Sound Technician to carry out at least basic maintenance. Equipment on OBs often has to be fixed on the spot and in a hurry because the crew is many miles from base and replacements cannot be obtained quickly.

Current Developments

Each of the tasks mentioned above has tended in the past to be a separate function and the Sound Technician progressed through most if not all over a period of years. Changes in technology and operational practice mean, however, that many of the functions are combining. In many companies there is a change in emphasis away from technical expertise towards greater creativity and artistic appreciation in sound.

There are exciting times ahead in television. Stereo television is being introduced and many companies are looking seriously at digital audio throughout the broadcast chain.

Careers Structure for Sound Technicians

Entry is normally at Trainee Sound Technician level with promotion to Sound Technician after nine months, followed by a further six years of automatic progression to the grade of substantive Sound Engineer. The period and grade may vary from company to company and early progression may be possible.

Thereafter, promotion is by merit and competition through two or more grades to a senior supervisory grade Sound Engineer in charge of the technical staff and facilities at, say, a studio, an OB unit, or a group of technical facilities departments. In most companies, Sound Engineers at substantive grade and above will be skilled in many or all of the jobs described above. Some companies, however, prefer their staff to specialise in a small number of skills.

Typical Recruitment Profile

Trainee Sound Technician (Video and Film)

	Essential	Desirable
Physical		
Age	18 +	21 +
Health	Good stamina	
Eyesight	Normal colour vision	
Hearing	Acute aural perception	
Coordination	Quick, calm, ear–hand reactions	
Speech	Clear	
Education and Training	Broad general education to at least GCSE (grades A to C) standard including Maths and Physics	(1) BTEC/City and Guilds/HND/ Degree in electronics, computer science, programme operations or communications or (2) BA Tonmeister or (3) Film and TV course at College of Further Education
Interests	Flair for some aspect of audio work, e.g., musical performance or	Wide interest in film and television. Electronics as a hobby

	Essential	*Desirable*
	recording, or public address work. The modern music scene. A knowledge of musical notation is essential for operational aspects of sound	
Personality	Equable temperament, able to work as a member of a team. A creative outlook on problems	Keenness to take initiative and responsibility
Availability	Able to work long unsocial hours including weekends at any location	

SPECIAL EFFECTS

The term 'special effects' covers an enormous range of techniques, carried out by many different people using anything from a piece of string to a computer. It can perhaps be described as the 'distortion of reality'.

Very few people indeed are employed full-time on special effects in ITV and they are only found in some of the largest companies. For many staff in TV, the creation of effects is a part of their job, but many effects are commissioned from outside companies which specialise in this field.

To most people, the term special effects brings to mind the flying sequences in the 'Superman' films or the space ships in 'Star Wars', but this is only a very small part of the story. Effects fall roughly into two categories: those that are physically built or made, and those that are generated electronically on a computer. The latter are being increasingly used in television, but the decision on which method to use will depend upon the degree of reality that can be achieved, on the cost and to a degree, on fashion. A company specialising in computer-generated effects, for example, turned down a request to create a sequence in which a pterodactyl was to fly around the Albert Hall. This could be achieved electronically, but it would not look as realistic as it would if a model pterodactyl was used. There are fashions for electronic effects, and indeed for reality. A television commercial which showed a car being driven through a field of burning crops was filmed in reality without any special effects.

Physical effects such as models, puppets and explosions, and 'optical' effects such as back projection and split mirrors are normally created by specialist companies working primarily for the film industry. There are probably no more than two or three people working full-time in this area in ITV. The work is enormously varied and is limited only by the imagination. The Programme Director and the Set Designer will describe what they want to achieve, and it is the task of the specialist to create that effect in the way that looks most realistic and unstaged within the budget allowed for the task. This calls for a great deal of ingenuity and persistence, trying things out again and again using different techniques and materials until it works. There are few rules to follow. Many techniques are passed on by word of mouth or by inquisitively working out how someone else has achieved an effect. You need to be pushy, cheeky, single-minded, and a determined perfectionist. Some things may be learned the hard way, such as how to create a house on fire without burning down the building! The use of fire and explosives (pyrotechnics) can be dangerous and there are certain rules on safety which must be followed. It is important to remember that in the area of 'physical' special effects, you may have responsibility for people's lives.

There is no clearly defined career path to follow. Some specialists have begun their careers as Electricians, or in the Props Department, both of which can provide relevant skills. A background in subjects as diverse as engineering and sculpture is useful, for example, in model making, and in puppet making. A knowledge of mechanics is useful if you are to 'rig up' a special effect such as a tyre bursting on a car. Practical skills are clearly important, but they will be of little value unless combined with creativity, artistic ability, and the skill of thinking laterally as well as logically. Many people tend to specialise in the use of a certain medium according to their particular skills and interests.

There is no set pattern of promotion in this field. You do the job because you love it, and are happy to spend your nights lying awake trying to work out how to overcome a problem.

Video effects, i.e. those which are created electronically by computer, are also made predominantly by specialist companies working outside the ITV network. They provide a service to the broadcast companies but also to the makers of television commercials, promotional videos, and many others. The service they offer is primarily at the 'post-production' stage. Once the programme has been recorded by an ITV company, for example, it is sent to the specialist organisation or 'facility house' to have the video effects added. Once again, it is difficult to describe adequately the kind of effects that can be produced since they are limited only by imagination and the capability of the computer. The Channel 4 logo and the introductory sequence to ITN's 'News at Ten' are examples of video effects as are most 'animated' sequences, i.e. drawings that move.

Such effects are created on a variety of computer graphics machines, most of which have slightly different capabilities. Some are capable of producing a three-dimensional effect, giving the impression of light, waves, ripples, etc. Many advertisements use video effects. Some incorporate video effects, physical effects and reality superimposed on each other, and involve cooperation between several different companies.

Some video effects are added 'live' to television programmes, for example the live pictures from a light entertainment show can be fed instantly down a communications line to a video effects company. The company offers appropriate effects to the Programme Director back in the control room. The effects appear on a screen in front of the Director and are added to the live programme as appropriate.

There is no set career path into video effects. Some would argue that a background in electronics engineering is useful in order to understand how the computer graphics equipment operates. Other employers look for computing specialists and will only accept people who can programme the computer using the appropriate language. In some cases the video effects specialist is required to programme the computer and in some cases not. Some have a background in the sciences, for example a degree in Physics, while others have a background in the Arts. It is not necessary to have a degree in any of the subjects mentioned but many people do, and the competition for jobs often means that a degree is an advantage. A logical mind is, however, vital and must be combined with a high level of creativity and artistic flair.

Jobs are rarely advertised. Those within the ITV companies are normally filled internally, while those in specialist facility companies are often filled by people who have started at the bottom as runners, or who have turned up on the doorstep at the right time.

We have looked so far at people who are employed full-time on special effects, but for many different staff in ITV, special effects form a part of their job. The Make-up Artists, for example, may be called upon to produce special effects such as ageing a character, or turning a man into a werewolf. The Vision Mixer will create video effects such as splitting the screen or spinning the picture, and the work of the Graphic Designer will incorporate many video effects using the same computer graphics machinery mentioned above. The Props Department may from time to time be called upon to build, for example, a magic castle for a children's programme. Video Editors, Lighting Directors, Costume Designers, and almost everyone involved in the creative aspects of programme-making will be concerned with special effects from time to time.

A career in special effects is exciting, challenging and possibly one of the most creative there is. As one specialist put it, it is 'living out everyone else's fantasy'.

No recruitment profile is given for this job.

STAGEHAND AND PROPS

Another very interesting area of work in television is stage and properties. Stagehands, or Setting Assistants, are the people who erect the scenery in the studio or on location after it has been made by the craftsmen. Their job is very similar to that of stagehands in the theatre. Scenery for the most part consists of large 'flats' made of plywood mounted on wooden battens. It has to be assembled piece by piece and is fastened together and supported by all manner of clips, ties, braces and struts. Stagehands are skilled at interpreting the Designer's floor plan and drawings in order to erect the sets in exactly the right part of the studio or location. They work in small teams under the supervision of a Scenemaster (or equivalent) putting up the set, working some special effects, captions and any scenery changes during the show, then dismantling and removing the scenery to store after the recording or transmission is over.

Camera crew and Props at work on location (YTV)

The Property staff, known as 'Props', do a similar job to the Stagehands except that they are concerned with all the 'action props' used by the artists, e.g., hot-air balloons, cutlery, crockery, food, drinks, cigarettes, telephones, pianos – the list is endless. In some companies the jobs of Stagehand and Props are combined.

Within the Properties Department there are also to be found Propmakers. These are the people who manufacture the models or fake structures needed for special visual effects. They work in a variety of material such as papier mâché and fibreglass to create realistic scale models of buildings, ships, machinery, in fact anything that the Writer's and Designer's imaginations may dream up. It should be noted, though, that the really spectacular special effects associated with space thrillers and James Bond films are normally the work of highly skilled specialists hired in for the production.

Both Stagehands' and Props' work is, for much of the time, very strenuous, requiring both physical strength and stamina. Long and irregular hours are worked and in most companies location work is a regular duty. The skills of a Stagehand can be acquired by any reasonably fit and intelligent person with the ability to learn to read drawings and tie fancy knots. Prophands' and Propmakers' vacancies are usually filled from the experienced Stagehands' ranks, when an artistic eye and/or craft skills may be sought. Some Stage- and Prophands go on to become Property Buyers which is a very specialised job.

In some companies a small number of Stage and Props staff are recruited with a view to their potential for promotion to floor management. Unless genuine opportunities for such advancement exist, a creatively ambitious person would be well advised to avoid seeking a Stagehand's job as a means of entering the industry. But for anyone who likes to be physically active and has a taste for, and preferably some experience of, behind-the-scenes stage work the life of a Stage- or Prophand can be very satisfying.

Typical Recruitment Profile

Trainee Stage- or Prophand

	Essential	Desirable
Physical		
Age	18 +	25–45
Health	Strong, fit, with good stamina	
Vision	Normal	
Hearing	Normal	
Coordination	Quick reactions	
Education and Training		GCSEs (grade C) in English, Maths and Technical Drawing. Clean driving licence
Interests	Stagework, music, any crafts	
Personality	Equable temperament, a good team member, outgoing, initiative, self-confidence	

	Essential	*Desirable*
Availability	Able to work long irregular hours and work away on location	

TRANSMISSION CONTROLLER/ PRESENTATION CONTROLLER

The work of a television station falls very broadly into two areas. On the one hand, there are the Producers, Directors, Camera Operators, Sound Technicians, Editors, etc., who make programmes which, in the main, are stored away for future use. On the other hand, there are the staff who are responsible for the station's output of programmes on that particular day. Some staff such as technicians and journalists are involved in both aspects of the station's work, but a small number of staff are concerned more or less exclusively with the day's schedule of programmes, and it is this aspect of television that we are concerned with here.

The more traditional role of Transmission Controllers is to send the output of programmes in sequence, and on time, to the local IBA (Independent Broadcasting Authority) transmitter. They are the last link between the station and the public and as such, have a vital role to play. The station's image with the public depends not only on the quality of the programmes but also on the smooth progression from programmes to commercials, to announcements, to news flashes, etc. It is the Transmission Controller's responsibility to ensure that this is done. If a Director makes a mistake while recording a programme, the scene can be re-shot, but if the Transmission Controller makes a mistake, the result is there for the viewing public to see.

The working environment of the Transmission Controller can appear somewhat daunting at first to the outsider. There is a large presentation mixer console covered with buttons, switches and lights, and a bank of television monitors showing such things as the picture going from the station to the transmitter, the picture going from the transmitter to the home, and the start of the next programme in sequence. The room is permanently darkened and the Transmission

Transmission Control (YTV)

Controller is unlikely to see daylight for many hours.

One of the most important pieces of equipment in the room is the clock, since the Transmission Controller must press the appropriate switches to bring in the next commercial, programme, or announcement on the second. Split-second timing is vital because of the network system by which ITV operates. Any one of the ITV stations might be the originator of a programme which is sent on cue via a land line or other form of link to the other stations. From there the Transmission Controller sends it on to the local transmitter. At the same time all commercials originate from the local station, as do local news programmes and announcements. Coordination is essential and if the Transmission Controller fails to react quickly, the region might miss, for example, the start of 'News at Ten'.

It follows that the job involves a considerable amount of planning and preparation before the start of a shift. The Transmission Controller must for example be thoroughly familiar with the planned programme schedule, where programmes will originate, which commercials are to be transmitted and their exact length, which back-up programmes are available in the event of a problem and if any news items are likely to cause a break in the planned schedule. A thorough knowledge of IBA rules on what can and cannot be transmitted is also essential.

The job combines hectic activity and pressure with periods of relative inactivity between commercial breaks. However, concentration is vital at all times in case something goes wrong. If there is a technical failure in the middle of a film, for example, it is the Transmission Controller who must ensure that the public is not left with a blank screen. It is a mixture of teamwork and working alone, of periods of intense stress and apparent inactivity.

The job requires an intelligent, quick-thinking approach combined with a high degree of manual dexterity. In addition, editorial perception and a good visual sense are needed. Many Transmission Controllers have been skilled typists, telegraphists, and sound mixers. They are quick to make decisions, have good spatial perception, speed of response – and a liking for the detail of administration.

Recruitment is usually at trainee level and promotion to Assistant Transmission Controller takes place after nine months. Promotion to Transmission Controller may take up to 10 years but this depends on ability and the number of vacancies. Many trainee vacancies are filled internally by staff from a variety of departments.

The role of the Transmission Controller is, however, changing rapidly with the introduction of new technology which enables many of the traditional tasks to be carried out automatically.

In many companies the Transmission Controller is assuming new responsibilities, and a new breed of 'Presentation Directors' is emerging. The Presentation Directors, in addition to transmission control duties, may also be involved in such things as the producing, directing, and scheduling of on-screen promotions for forthcoming programmes.

Typical Recruitment Profile

Trainee Assistant Transmission Controller

	Essential	Desirable
Physical		
Age	18 +	21 +

	Essential	*Desirable*
Health	Good	
Vision	Normal (able to look at television monitors for long periods)	
Hearing	Good	
Coordination	Excellent hand–eye coordination. Quick reactions	
Speech	Good, clear speech	
Education and Training	Broad general education to at least GCSE (grades A to C) standard	'A' levels or degree (any subject)
Interests	Interest in television. Good knowledge of current affairs and well informed on the broad ranges of topics covered by television programmes	
Personality	High degree of concentration. Social skills. Ability to work on own initiative under pressure. Equable temperament. Leadership	
Availability	Unsocial hours. Transmission Controllers are on shifts covering 24 hours per day, 365 days per year	

VISION MIXER

The Vision Mixer's job is to assemble sequences of visual images from various sources during the making of a television programme. It is a job which combines operating skills with artistic interpretation and is a vital part of the production process.

The pictures which make up a programme come from a variety of sources; for example, from cameras in the studio, pre-recorded video tape, from a 'telecine' machine (which transfers film into a television picture), or from slide photographs. All of these sources are available to the Programme Director who must decide which ones to use, and in what order.

If we look at the example of a news programme, the Programme Director may decide to start with a shot of the main Presenter in the studio on camera 1, then cut to a shot of the second Presenter on camera 2, followed by a video tape illustrating the news, and finally a slide photograph of a person associated with the story. It is the task of the Vision Mixer to make the cuts from one source to the next at the appropriate moment, on receiving a cue from the Programme Director. The job is in essence 'live' picture editing.

The cuts are made on a complex electronic machine known as the vision mixing console which is located in the studio control room. All the vision sources are fed into the console and as the Vision Mixer operates the controls the end result is recorded on to video tape, or transmitted live according to the programme.

The vision mixing console can also produce special effects to make the transition from one source to another more interesting. It can, for example, fade one picture into the next or 'wipe' from one picture to another. The complexity of consoles varies considerably, but some are capable of digital effects such as spinning the picture around, or dividing the screen into a pattern of small pictures. This can produce an attractive result on a light entertainment show for example.

Vision Mixers also work on outside broadcasts such as church services, sports events, etc. Each outside broadcast vehicle is equipped with a vision mixing console, usually with fairly limited capabilities. However, not all OBs require the services of a Vision Mixer.

A good Vision Mixer will be able to anticipate when the Director will give the cue to make the cut. This is important because the exact moment at which the cut is made can be critical, and a slow reaction to a cue can ruin the desired effect. A badly timed cut can spoil the mood of a sequence. Similarly in musical programmes it is important that the cut is made exactly on the right beat. Very quick reactions, artistic judgement, a sense of rhythm and a feel for music are therefore essential.

Although the modern vision mixing console is a very sophisticated piece of equipment it is not necessary to be an electronics expert in order to operate it. It does, however, employ advanced digital electronic techniques and it can be an advantage to develop a feel for the logic in order to get the maximum performance from the machine.

Vision Mixers spend most of their working day in the control room which is dark. They must be able to concentrate on television monitors for long periods, carry ideas in their heads, and follow instructions quickly and accurately. They must also be able to work under considerable physical and mental pressure from time to time. Manual dexterity is important in order to locate and press the correct buttons on the console.

Trainee Vision Mixers are usually recruited from among the existing staff of the television companies. No particular background experience is specified and applicants come from a wide variety of jobs in television including technical, administrative and secretarial. Very occasionally, trainee Vision Mixers may be recruited externally, but this is the exception rather than the rule.

Typical Recruitment Profile

Trainee Vision Mixer

	Essential	Desirable
Physical		
Age	18 +	21 +
Vision	Good colour vision	
Hearing	Good (able to hear instructions clearly)	
Coordination	Excellent hand–eye and hand–ear coordination	
Education and	Broad, general	Film and/or

	Essential	Desirable
Training	education to at least GCSE grades A to C	television production course. Basic knowledge of geometry is an advantage in operating digital video effects
Experience		Experience of working in some aspects of television is highly desirable. Able to read musical scores
Interests		Television, theatre, amateur dramatics, music, visual arts
Personality	Quick reactions and a high degree of manual dexterity. Able to follow instructions quickly and accurately. High degree of concentration. Able to work as a member of a team. Able to use initiative and keep calm under pressure for long periods. Artistic and musical appreciation	
Availability	Able to work irregular hours occasionally away from base	

WRITER

Television Scriptwriters are normally employed on a freelance basis for a particular programme or series, and are only very rarely permanent employees of a television company. The education, training and experience of Writers is very varied as there is no set career path to follow, and it is therefore only possible to provide very general information here.

A sound educational background, especially in literary subjects such as English and possibly foreign literature, is a firm basis from which to begin, but no specific qualifications are required. Many universities and colleges of higher education offer courses in creative writing, either as part of a broader full-time course or as specific short courses. There are also evening classes both in general writing and specifically for radio, film and television. Details of such courses can be obtained from the universities and colleges, local arts associations and education establishments. You may also find details at your local library of writing circles where people who are interested in writing can obtain advice and criticism – essential ingredients in all creative work.

A lucky few may break into television scriptwriting by writing an original piece and sending it unsolicited to television companies. Others may prefer to seek out a reputable literary agent who can find a suitable customer for their work. Agents will of course require fees for their services.

Fresh, new ideas are always very welcome, but potential Scriptwriters should also bear in mind that in order to earn a living they may have to respond to other people's ideas. They may, for example, be asked to write a script for a long-running soap opera. This will mean creating new situations for well-established characters in a style which blends in with that of previous writers. Other Writers may be employed to adapt well-known books into a format which is suitable for television.

Comedy scriptwriting is a specialised skill, and established comedy writers are commissioned to write scripts for series in much the same way as drama scriptwriters. It is, however, possible to begin by submitting single jokes for consideration by specific comedians. It is clearly necessary to tailor the joke to the particular style of the comedian. Some comedy scriptwriters are lucky enough to begin their careers by having complete programme scripts accepted by a television company, but this is comparatively rare.

Scriptwriting is a very speculative profession and often very insecure. It is subject to current fashions and topical demands, and successful writers will anticipate these rather than merely write what appeals to them. It is also very important to remember that scripts must be suitable for the visual medium. Writers often have to face numerous rejections and strong criticism which can be very discouraging. Perseverance and tenacity are essential characteristics as well as creativity and writing skills.

No recruitment profile is given for this job.

OTHER JOBS

There are a number of jobs within television which are highly specialised and which are rarely available as direct entry points to the industry. It is also very difficult to advise on the kind of background that is needed for these posts. Luck, chance, and being in the right place at the right time are the important ingredients.

Casting

Not all of the ITV companies have a Casting Department, but those that do normally employ Casting Directors at a senior level and Booking Assistants at the junior level (although titles may vary).

Prior experience in the acting profession is usually required. Typical previous experience might be employment in a large theatrical agency.

Continuity

The job of Continuity does not exist as a separate entity in television as it does in the film industry. Responsibility for ensuring that the visual flow from one scene to the next is correct lies with the Production Assistant, but forms only a small part of the job.

Grips

These are specialists who rig scaffolding and camera rails, special camera cranes and mountings. The job title comes from the film industry and where TV productions require them they are generally hired in from specialist companies. Some ITV companies now train their own but they are always recruited internally, normally from Stagehands and Rigger/Drivers. The use of Portable Single Camera (PSC) has led to more location work and so to a few more opportunities for Grips in ITV.

Production Buyers

Production Buyers or Property Buyers are responsible for purchasing items needed for inclusion in programmes. They work closely with the Set Designers in order to purchase appropriate furniture, food, greenery, books, or anything else that is needed as a prop. They rely heavily on personal contacts, so experience of the industry is essential. Entry to the job tends to be either from the Props Department, or from theatrical stage management.

Programme Manager/Production or Technical Coordinator/Technical Supervisor/ Production Manager

The range of job titles indicates what a wide variety of tasks this kind of job can cover. Duties can include acting as a general 'trouble shooter' and coordinator during the making of a programme, or planning the logistics of how the programme can be made. These jobs vary considerably from company to company. All require extensive knowledge of TV production.

Sports Associates

These are usually experienced television Journalists who also have a deep interest in sport and have been lucky enough to be able to combine the two. Very few people are employed in these posts.

Weather Reporter

Most of the Weather Reporters who are responsible for the main weather forecasts are qualified meteorologists who came into contact with television through their work with the Meteorological Office. Some, however, are personalities who have been identified by the television company as having the right qualities to present the weather forecast during, for example, a magazine programme.